THE Process-Centered Enterprise

The Power of Commitments

GABRIEL A. PALL

Forewords by

Dr. A. Blanton Godfrey
Stephan H. Haeckel

St. Lucie Press
Boca Raton London New York Washington, D.C.

Library of Congress Cataloging-in-Publication Data

Pall, Gabriel A
The process-centered enterprise : the power of commitments / by Gabriel A. Pall.
p. cm.
Includes bibliographical references and index.
ISBN 1-57444-239-2 (alk. paper)
1. Industrial management. 2. Customer services. 3. Quality of products. 4. Business.
I. Title.
HD31.P26 1999
658.4′063—dc21 99-048848
CIP

Direct all inquiries to CRC Press LLC, 2000 N.W. Corporate Blvd., Boca Raton, Florida 33431.

St. Lucie Press is an imprint of CRC Press LLC

International Standard Book Number 1-57444-239-2
Library of Congress Card Number 99-048848
Printed in the United States of America 2 3 4 5 6 7 8 9 0
Printed on acid-free paper

Foreword

A few years ago, one of the hottest trends in business in the U.S. was reengineering. Some companies even had "reengineering departments" and "directors of reengineering." Courses on reengineering were packed, and it was almost impossible to pick up a magazine or journal that did not have at least one article on reengineering in some form or another. And then, almost as suddenly as it had appeared on the scene, it disappeared. Even some of the best proponents of reengineering were writing articles about its failures and why it didn't work. Now, in some companies, the term "reengineering" is no longer allowed to be spoken. It has become so associated with downsizing that it creates noticeable panic whenever mentioned.

Other companies have quietly integrated reengineering into their total quality system and moved on to other areas of emphasis. They have found many of the concepts to be valid and useful and continuously apply them as they strive to improve organizational performance. Some organizations have even found these concepts have fundamentally changed the way they think about their business operations. The question has been how to separate the wheat from the chaff. There were many great ideas and new practices which came with the business process reengineering flurry. But there was also a lot of hype and unfulfilled promises.

The true pioneers of business process quality management — or business process quality improvement, as it was first called — included companies such as IBM, Ford, AT&T, and NCR. These companies took basic quality improvement and control ideas and applied them where no one had ventured before, to the administrative and business processes. Most of these ideas had been applied for years in the production processes, but many of the business processes were virgin territory.

Some of the early results were truly staggering. Cycle times were reduced by over 90%, with reductions in staff of 50% or more. Processes were simplified from 20 or 30 steps to three or four, or even to one. Paper flows were cut drastically, and sometimes paper forms were eliminated completely. Insurance agents were completing claims at auto crash sites; orders were being taken, filled, and shipped in a few minutes vs. weeks or even months; customers were able to check order status on line rather than going through four or five customer representatives.

But what went wrong? With so much promise and such potential, why did so many companies fail? There have been numerous articles (and even a few books) about this, but basically there have been a few clear reasons. One reason was the endless search for the "silver bullet" — the one method that would solve all problems. Many managers failed to understand that process thinking was a fundamental new idea for them, and they failed to understand the critical differences between task management vs. process management. Without the fundamental ideas of quality management being well understood and practiced within an organization, the organization does not have the tools to manage the critical processes.

A second reason was the tendency to try to "buy" reengineering. Hiring a strike force of outsiders to come into one's organization to quickly destroy the old processes and instantly install new, effective, efficient processes sounds far better than it works. Our own people still have to manage these processes in the long run. They have to make the measurements, fix the problems, and continuously improve both efficiency and effectiveness. The third reason was the failure to understand the tight connection between truly world-class business processes and information technology.

One of the true pioneers in business process reengineering was Gabriel Pall, formerly the director of IBM's Quality Institute and now one of my colleagues at Juran Institute. Gabe's book, *Quality Process Management,* published in 1987, was the first significant book on the subject and the one that introduced me — and many others — to these concepts.

In his new book, *The Process-Centered Enterprise: The Power of Commitments,* Gabe shares with us his views on the two major developments in business process reengineering in the past five years.

First is the profound recognition that a business process is not just the transformation of inputs into outputs but also comprises — and is driven by — the communication and management of commitments. The second development is how much the various reengineering methodologies have been enhanced by their integration with advanced information technology tools and systems. Software tools such as ABC Flowcharter and Visio have greatly aided process design, and integrated enterprise applications systems such as Baan and SAP have enabled the simultaneous redesign of the process flow, human resources/organization, and their enabling technologies.

For many companies reengineering has become a valuable tool in their continuing efforts to build world-class organizations. Used intelligently, these ideas lead to major breakthroughs in business performance.

*Dr. A. Blanton Godfrey**

* Dr. A. Blanton Godfrey is Chairman and CEO of Juran Institute, Inc. He has been involved in business process reengineering ever since he led a benchmarking team to study IBM's business process management while he was Head of Quality Theory and Technology at Bell Laboratories. AT&T was soon implementing their own version of "business process quality improvement."

Foreword

What is the future of process design? One thing is sure: It will not be more of the same. What got us here, in terms of progress in business process improvement, was a conviction (with results to back it up) that designing repeatable and predictable workflows could substantially improve business performance. "Starts here; ends here," "input, process, output," "defect detection and correction," to mention but a few of the process management mantras of the 1970s and '80s, is language that betrays both the industrial age and information technology (IT) professional origins of this conviction. The pre-information age managerial premise of determinism and an ability to control one's corporate destiny was very consistent with the IT professional's premise that it was both necessary and possible to reduce processes to procedures. Procedures can be automated, and automation promised the ultimate efficiency in doing what it had been determined should be done.

But, in the present and future environment of increasing unpredictability about what customers will want, and therefore what it makes sense to do, businesses can no longer rely on predefined processes that require predictable inputs. What, then, can and should they rely on? And what is the future of process design if there is no future in predictability about what the firm will be called upon to do?

The answers to these questions lie in distinguishing process and procedure. By their very nature, *procedures* deal only with contingencies that have been thought of in advance and do not deal with unanticipated inputs. But skilled humans often can. That's the key: It takes people, not procedures, to deal with the unexpected. This is why, as the business environment becomes less and less predictable, capturing the *ad hoc* dimension of processes becomes more and more important. In fact, many highly adaptive *processes* are almost exclusively *ad hoc*: the creative process of starting up a new business or the process of raising a first child, for example. At the other extreme is a process such as renewing a driver's license, which is almost entirely procedure. Occupying the middle ground are some high-performance groups, such as surgical teams, cockpit crews, or special forces units that carry out processes with high degrees of both procedural and *ad hoc* activity.

Doing the wrong thing better and faster is an obviously bad idea, yet it is the implicit objective of many process improvement projects. Confusing rapid response with adaptiveness, managers sometimes make the dangerous assumption that the way to deal with unpredictable change is to identify, codify, and reduce the execution time of their core processes. While this is an excellent recipe for efficiency and competitiveness when things are predictable, following repeatable, definable procedures in times of discontinuous change is at best useless and at worst disastrous. In these circumstances, not only must core processes change dynamically, but the designation of what is "core" must frequently change, as well.

Adaptiveness implies appropriate response to change, not just rapid response. If the change is unanticipated, then the response will often have to be unprecedented. This means that for large organizations to adapt systematically in the face of the unexpected, they must create and codify *ad hoc* processes, in addition to the procedures that have been designed in advance to deal with those things that change incrementally and with relative predictability. But *ad hoc* process design poses a significantly different challenge than most process experts deal with. A procedural design answers the question, "Who (or what) does what with what to what in what sequence?" An *ad hoc* design, on the other hand, answers the question, "Who (never 'what') owes what to whom?" *Ad hoc* process designs codify the outcomes, or *ends,* of business processes and specify the producer and supplier roles for those outcomes; procedures codify the activities, or *means.*

In *The Process-Centered Enterprise: The Power of Commitments,* Gabriel Pall brings the idea of adaptive process design to a discipline steeped in the procedural tradition. His rare combination of theoretical knowledge and years of practical experience helping companies improve their performance by improving their processes makes him especially qualified to address the challenge adaptability poses to process managers.

Pall and I offer different, but complementary, prescriptions for making large organizations adaptive. His primary emphasis is on process design, while mine is on systems design. In my view, processes are organizational capabilities — capabilities with the potential of producing outcomes. How these outcomes are linked to one another and how they achieve organizational purpose is a systems design issue rather than a process design issue. Systems designs are ends oriented; that is, they start with organizational purpose and decompose it successively into multiple layers of subsystem outcomes. Systems designs say nothing about activities or tasks. The management of organizations as systems therefore involves managing the interactions, rather than the actions, of subsystems.

Process design, on the other hand, is a means-oriented approach to business design. Processes are the means by which organizational capabilities produce their outcomes. If the processes are designed "from the top down," as Pall recommends, and if their starting point is the result, rather than the input, process and systems designs may produce similar, or at least compatible, results. Of course, if process design is bottom-up (as is too often the case), and if management emphasizes process excellence over outcome relevance, then the chances of getting better at doing the wrong thing increase significantly.

Notwithstanding the differences between the system and process approaches to organizational design, both Pall and I base our prescriptions for codifying adaptive behavior on the seminal work of Fernando Flores and Allan Scherr. The profound insight of Flores was that conversations between people about accomplishing something are dialogues about *commitments* between the suppliers and users of *deliverables*. Scherr showed how Flores' ideas could be used to represent process designs in a format that incorporates and sequences both the procedural and *ad hoc* dimensions of the process. An important result is that the generation and completion of spontaneous commitments between two individuals can be rigorously codified as a generic process. The execution status of the process can be tracked simply by determining the nature of the last communication between the parties. Because of the rigor of Flores' schema, and Scherr's modification of it, technology can be used to do the tracking.

This, in turn, makes it possible to codify the way empowered people in large organizations figure out, on the fly, how to define, negotiate, produce, and assess the deliverables required. Furthermore, technology can ensure that this creativity is enabled within the context of an overall business design that specifies how the processes are linked to achieve organizational goals. It is this attribute that reconciles process design with adaptive, *ad hoc* behavior. So, the theoretical underpinning of how to design processes for adaptiveness has been developed and its validity confirmed in practice. Now comes the difficult part: learning how to implement it well in practice. In this book, a well-crafted contribution to the process design literature, Pall launches professionals and managers up the learning curve of good adaptive designs. In doing so, he makes a very important contribution, because adaptiveness is the future of process design.

*Stephan H. Haeckel**

---

* Stephan H. Haeckel is Director of Strategic Studies at IBM's Advanced Business Institute and Chairman of the Marketing Sciences Institute. His book *Adaptive Enterprise: Creating and Leading "Sense-and-Respond" Organizations* (Harvard Business School Press, 1999) contains his prescription for systematic adaptiveness by large organizations. Haeckel's focus is on the new forms of strategy, organization, and governance that will replace the planning processes, hierarchical structures, and command-and-control management systems of the industrial age. It thus complements the process-centric focus of this book.

The Author

Gabriel A. Pall specializes in the design and management of business processes; the organizational, financial, and general management implications of company-wide continuous quality improvement; and the role of quality in strategic planning and competitive leadership. He is an internationally recognized practitioner of the quality of management: strategic business planning, change management, the management of technology, and business transformation. His functional experience includes law, finance, and information technology; his industrial applications expertise covers the chemical, pharmaceutical, paper, and manufacturing industries; higher education; and professional service firms. Prior to joining Juran Institute as a Senior Vice President, Mr. Pall was Director of the IBM Quality Institute.

Mr. Pall is the author of *Quality Process Management* (Prentice-Hall, 1987) and *Introduction to Scientific Computing* (Appleton-Century-Crofts, 1971) and co-editor of *The ASTD Technical and Skills Training Handbook* (McGraw-Hill, 1995). Mr. Pall holds a B.S. degree in Structural Engineering from the Technical University of Budapest, a Civil Engineering degree from the Technical University of Vienna, and an M.S. degree in Engineering Mechanics from the University of Pennsylvania.

Mr. Pall and his family live in Stamford, Connecticut.

Acknowledgments

My views on processes and business process management have been shaped by three factors over the years:

- First, my earlier association with the IBM Quality Institute and then with the Juran Institute, where I was able to pursue the study and development of business process management concepts and methodologies as my professional and personal priority.
- Second, leading the practical implementation of numerous business process design and reengineering projects in the United States and abroad, of which the Union Camp and Pfizer cases are only the most recent examples.
- Third, observing and learning about the work of, and especially through direct contact with, those who have provided leadership, through examples and writings on key areas of process management: Ackoff, von Bertalanffy, Flores, Haeckel, Hammer, Juran, Keen, Scherr, and others.

It is very difficult to give proper credit to all the people who have assisted me in the preparation of the manuscript or who helped me in one way or another to make this book a reality. To thank everyone in the manner they properly deserve would require another chapter. However, I do want to take this opportunity to acknowledge the efforts of those who contributed most: Drew Gierman of CRC Press, for his confidence and initial encouragement; Leslie Kelly, who helped define and position the book and also made invaluable comments and editorial suggestions regarding its contents and structure; and Blan Godfrey of Juran Institute and Steve Haeckel of IBM, for their very appropriate and timely forewords.

I owe sincere thanks to Barb Allen, Joe Brenner, Les Crandall, Vic Haburchak, Stafford Kelly, Paul Kleindorfer, Jim Riley, Kent Steele, Rob Street, and Audrey Woods for taking time from their very busy corporate and professional schedules to review the manuscript. I would also like to say special thanks to Marilyn Maher and Carole Wesolowski for their help in processing and organizing the text; to Debbie Hurd and Karen Clauser, who spent a tremendous amount of time on the editorial tasks connected with the book; and to Jane Gallagher and Laura Sutherland, for their graphics work of unparalleled quality. Last, but not least, I am indebted to the Juran

Institute for its generosity in providing copyrighted materials and administrative facilities to help my efforts.

To all of these people and many others, I wish to express my personal gratitude.

Introduction

Not a single corporation exists in the developed world today whose management does not want an organization flexible enough to respond quickly to changing market conditions, lean enough to beat any competitor's price, innovative enough to deliver its products and services in a technologically superior way, and committed enough to deliver the highest quality to its customers. But, if corporate management wants their organizations to be all this, why are so many companies still refusing to change, ignorant of customer needs and, ultimately, not profitable? The answer lies in:

- How these organizations do their work
- How they manage the decisions and commitments that drive it
- Why they cannot effectively respond to unexpected change

The Business Need for Adaptability

Responding to change became the most important competitive issue in business and industry in the late 1990s and is growing in importance every day. In a recent series of surveys[1] of executives and heads of key business units of major North American corporations, over 50% of the participants identified accelerating and unpredictable change — also known as continuous discontinuity — as the number one management concern about the future. A study of 300 CEOs conducted in 1998[2] showed globalization and knowledge management as top executives' two most frequent worries. Clearly, the ability to sense change and capture knowledge about it is a key component of this. A common theme is now emerging regarding how the very best companies are responding to this challenge:

- Shift from the product to the process view.
- Empower more of the front-line people.
- Increase customer-related knowledge.
- Recognize the contribution of intellectual assets to productivity.

These four elements are not mutually exclusive but in combination represent a powerful approach to doing business in any industry, including the public sector.

Business management in the 21st century will be about responding to accelerating and unpredictable change through the rapid adaptation of commitments made to customers, the related organization of work, and the deployment of needed assets in the enterprise's processes.

With an increasingly global business environment and accelerating change, a number of significant economic and business developments underline the need for ongoing, effective, and rapid organizational response to change:

- In the 1990s, there has been an unprecedented worldwide growth in market capitalization as witnessed by the phenomenal performance of stock markets worldwide. By the end of 1998, Microsoft became the world's "most valuable" company with $267 billion market capitalization; the world's five highest-valued corporations now well exceed $1 trillion in market capitalization.[3] Paradoxically, however, these highest-valued companies are by far not ranked at the top of materials-based asset ownership list. The message is that the market has come to recognize the increasingly important contribution of the network economy, related technologies, and intellectual assets to value creation.
- The European common currency, the *euro,* is now a reality, with a strong integrating effect on worldwide finance, trade, and market development.
- A new, sweeping change is taking place in the "information economy." In the developed world, information technology now accounts for a growing share of gross domestic product (GDP). Worldwide, business-to-business electronic commerce is expected to reach $1.3 trillion by 2003, over 9% of total business-to-business commerce.[4]
- In the U.S., productivity growth again started accelerating in 1996, even though it was well into a long cycle of economic expansion. This somewhat unusual phenomenon may be the long-awaited effect on output growth of three decades' steadily rising investment in information technology, communications, and the intensifying "information economy."[5]
- Industrial and business leadership continues to shift to information technology companies; Microsoft, Intel, AOL, General Electric, Nokia, SAP, Oracle, and Cisco Systems have emerged as the companies of the 21st century.
- Ironically, the technology giants of the 1980s — IBM, AT&T, Siemens, Xerox, and Philips — are still working hard to reassert themselves and maintain a more competitive business posture in the face of accelerating, diversified, and relentless global competition.
- The automobile manufacturers of the world continue on a path of worldwide consolidation as exhibited by the merger of Chrysler and Daimler-Benz. Once-gilded automakers such as Jaguar and Rolls-Royce are now mere subsidiaries of their larger parent corporations. The 21st century will see probably no more than half a dozen giant automotive companies establish themselves in unassailable positions of competitive leadership.

- Total quality management (TQM) is now viewed as a given rather than a conscious priority. Where it was adopted, it is now another management approach embedded in the organization's everyday way of life.
- The wave of resource reduction and cost cutting (also known as "downsizing" or "rightsizing") came and, for the most part, went by the late 1990s. It even hijacked the term "reengineering" and gave an otherwise perfectly sound and appropriate management approach a bad name and feared reputation. Progressive organizations have refocused on growth and superior competitive position by means of more effective response to change.

In view of this, managers and professionals who want to be prepared for the new century must reject the narrow view that treats the response to change as primarily an issue of change management. Traditional change management has always focused on just the organizational component — that is, behavior, structure, and culture. If companies want to become competitive winners, they will have to look at how they get their work done and how they should adapt what they are doing to produce new outcomes when faced with sudden and unexpected change.

The basic premise of this book is that, to respond to and manage change, all levels in a company must take a process view focusing on all the business processes that make up the enterprise. Individually and cumulatively, these processes must have maximum adaptability, which will then enable the enterprise to do business differently — not just accommodate change, but leverage it for better results and further growth.

An enterprise behaving this way is said to be *process centered.* The process-centered management model that is introduced in this book has evolved over the past few years as the author has worked on a series of business process redesign projects, in a number of different kinds of organizations of various sizes in a variety of industries. During these projects, it became clear that while considerable time and resources were invested to achieve new and even ambitious levels of current process effectiveness and efficiency, the processes were not made demonstrably more adaptable and responsive to change. In the author's judgment, this is because management is still confused and apprehensive about the:

- Definition of process adaptability
- Design characteristics of an adaptable process
- Effectiveness of enterprise-level response that can be expected in the end

Based on example after example, our prescription for success rests on three premises:

- Companies organized by functional differentiation and managed hierarchically can no longer respond adequately to accelerating change in today's business environment and will, sooner or later, fail to stay viable.
- Companies that have reengineered some of their high-value, key business processes and streamlined their organization for better operating effectiveness and efficiency will be able to respond to predictable change.

- Only process-centered companies that have been able to reengineer all their high-value, key business processes for maximum adaptability and align their organizational structure around these processes will be able to respond effectively to both predictable and unpredictable change.[6]

A New Paradigm for Process Centering

Process centering is the pervasive business management approach to deal with accelerating and unpredictable change. It is a significant, but logical extension of traditional methods of process management and, in particular, process reengineering. Process centering is the leveraging and integrating mechanism that aligns commitments with the work activities and assets needed to meet them. It is also the prerequisite to the creation of an integrated value chain: the three-way information partnership between the enterprise and its suppliers and customers, the essence of *e*-business.

Beyond the issue of sudden and unexpected change, companies must continuously adapt regardless of external stimuli. To stay competitive, they must anticipate the future, innovate to better serve their customers, and understand the profound impact new and impending technology will have on their business. Added to this is the need to continuously hire, train, develop, and retain the best employees, because people run and operate the enterprise.

Executives and managers who are successful in the long term will understand their businesses, especially their key business processes, in great detail. As a result, they will also understand that any effective response to change must be anchored in a process-centered management approach, characterized by consistent decisions, commitments, and actions, which have to be communicated in an organized and disciplined way. While this concept sounds simple enough, its implementation can be quite challenging, involving all of the organization's business processes. Any executive or manager who is looking for a fast solution or the latest management fad will probably not find it in this book.

This book is aimed at the very essence of what a company or organization does and how it does it — its business processes. It encourages a top-down, process-centered approach that manages work, people, and technology by managing the commitments that drive them. To accomplish this, we first examine the five key requirements of process centering and then introduce process-centered management that makes them a reality.

Requirement One: Process Thinking

The first key requirement of the process-centering paradigm is process thinking, the recognition that in the face of change, the fundamental management construct is the business process, not the functionally oriented hierarchical organization. This leads to the formal definition of business processes and the remapping of the enterprise in terms of work performed in its various processes. Hammer and Champy's 1993 book, *Reengineering the Corporation,*[7] has shown the obsolescence of functional differentiation and provided the most persuasive argument for the need and importance of process thinking and the reintegration of enterprise work around its business processes.

Requirement Two: Process Adaptability

One cannot center an organization around rigid, inflexible business processes and expect it to behave adaptively; therefore, the second key requirement of the paradigm is the adaptability of individual processes, which collectively will result in an enterprise's ability to respond to change. Process adaptability means to respond to unexpected but detectable events by innovative changes and enhancements to the capability of the business process in question. Adaptability can be designed into business processes by creating the information flow and knowledge base that will enable pre-planned and *ad hoc* work activities to deliver the desired outcome. This, in turn, can be done by taking the holistic approach, which views the process as a chain of customer-supplier relationships comprising both commitment management and procedural workflow. It is the decision-making that leads to commitments in the face of change and the ensuing management of those commitments that fully integrate the human element into the business process. This logic leads us to clearly differentiate among (1) procedures and real processes, (2) the static and the dynamic, and (3) the predictable (deterministic), where a clear sequence of cause-effect relationships exists, and the unpredictable (probablistic), where the probability of something happening can only be estimated. We must also treat business processes as a series of commitments to customers, rather than mere work flow, which has been the traditional view of total quality management and conventional business process reengineering.

Requirement Three: Management of Commitments

Conducting business is essentially the negotiation and subsequent fulfillment of commitments between suppliers and customers along the value chain. In plain English, conducting business means:

- The conveyance and ongoing management (reviewing, updating, or renegotiating) of commitments between supplier and customer
- The management of work activities driven by those commitments

In business process management, managing commitments is the driving force; together with the management of work activities and procedures, it guides the complete execution of a business process.

Requirement Four: Information Technology

The fourth requirement of the paradigm is the expediting and integrating effect of information technology. This is a pre-condition of making adaptability happen because, in the face of change, rapid communication along the customer-supplier value chain is essential for making commitments and for a timely delivery on those commitments. Communication has to be clear, disciplined, and concise so that the people understand what has been agreed and committed to — especially in *e*-business, which is essentially the networked management of commitments along the value chain.

Requirement Five: Alignment

The fifth key requirement of the paradigm is the alignment of the organization around its key, adaptable business processes based on their value contribution to overall business performance. This is the only way for the enterprise to stay responsive to relentless change.

This process-centered model is now emerging as the dominant organizational form for the 21st century. Peter Keen's 1997 book, *The Process Edge*,[8] helped introduce the case for selecting business processes for reengineering on the basis of their economic value added (EVA) to the enterprise. This approach helps avoid the process paradox (discussed in Chapter 2): optimizing at the individual process level, but contributing little or nothing to the overall business performance of the enterprise. EVA has become a measure of financial performance and efficiency of business processes, which is also the mainstream of corporate financial management today. This reliable economic indicator has been missing from the fields of traditional business process reengineering and information technology management. And, because business processes are technology intensive and also represent a major portion of the capital assets of the enterprise, the model has to be based on the economics of capital. Information technology, a key component of business processes, has for too long been viewed as a mere support function and an expense item with its capital dynamics and value contribution largely overlooked.

The same thing goes for the information content of business processes. They have been viewed primarily as embodiments of work flow and the deployment of work teams, but certainly not as tangible corporate intellectual and knowledge assets.

Preview of the Book

This book is not a tutorial, but rather a guide to implement an advanced, logical approach to the business. Both a derivative and enhancement of my *Quality Process Management* (1987), it is designed for anyone who faces the managerial tasks of designing business processes which will be highly adaptable and responsive to change and establishing a process-centered model for their organization, aligned around its adaptable business processes.

The approach described in this book has been implemented on an industrial scale. Case examples, practical illustrations, and references are included; therefore, it should also help educators and consultants structure courses and training events that address the fundamentals of business processes, adaptable process design, and effective response to change.

Case Examples

The book relies heavily upon examples of both successful and failed cases of the management approaches discussed. Most examples are described briefly; others are explained at some length. The latter are drawn from field work, most of it conducted personally by the author. Among the case examples discussed in more detail, three deserve mention here at the outset: International Paper/Union Camp's Fine Paper Division, Pfizer Central Research Division, and IBM Austin.

The case of the order-fulfillment process of Union Camp's Fine Paper Division illustrates the reengineering of a critical core business process and its enterprise resource planning (ERP) system implementation. As Union Camp management realized that competitive advantage could no longer be maintained through high product quality or aggressive pricing, it looked to the responsiveness and speed of order delivery as the critical factor. This recognition led to the selection of the order-fulfillment process to be reengineered, primarily for unprecedented speed in customer response and near-perfection in on-time delivery of products. To ensure measurable attainment of these ambitious goals, a Baan enterprise system solution was integrated with the process redesign.

The reengineering of the procurement process of Pfizer's Central Research Division provides insight into two outcomes of the redesign effort. First, the functionally compartmentalized purchasing operation was redesigned into a homogeneous, overarching process to provide direct service to thousands of research scientists and knowledge workers. Second, through the expedient acquisition of the supplies and services necessary to feed the creativity of knowledge workers, the new process has become an example of process-leveraged organizational learning.

The new product introduction process at IBM's manufacturing plant in Austin, TX, was among the first industrial-scale implementations of the holistic process model and commitment management, the focus of this book. Since its introduction, this methodology has been incorporated into the plant's business process improvement program and has revolutionized IBM Austin's approach to process definition, analysis, and management.

Structure

The material is structured to lead the business planner and implementer through all the key concepts as well as some of the practical steps required to develop a process-centered enterprise structure that will truly be responsive in the face of change. To accomplish this, the book is divided into three parts.

Part I

This part explores the factors that provide the enterprise with the advantages needed to be more competitive. It introduces the fundamental concepts and rationale related to process centering. Also, because there is still considerable confusion and disagreement over the key components and performance characteristics of business processes, a set of rigorous definitions is presented. It seems that too many experts in too many books and courses have presented their own version, terminology, and approach to a fundamentally common theme. Yet, for the successful management of response to change, the conceptual and practical solutions can and must be clearly defined.

Part I examines the nature of change and traditional ways of coping with it. A discussion of process thinking and business process basics is followed by the introduction of process adaptability as the central requirement for responding effectively to change. The major attributes of process centering as a business transformation are presented, along with its key components: humans as empowered, accountable individuals, team-based management structure, and networked information technology. These, in the aggregate, make up the process-centered enterprise model, competitively superior because it is maximally responsive to change.

Part II

This part applies, in a systematic and organized way, the concepts and methodologies of process centering within the enterprise model described in Part I. Key to this management approach is the holistic process model that is based on the premises that all process work is (1) driven by commitments made to the customers of the process, and (2) supported by a chain of subordinate commitments within the process itself. In the face of change, the required outcome of process work will have to change, which means that work itself, and how it is organized, will also have to change. Ultimately, this means changing (or, more appropriately, renegotiating), reprioritizing, and restructuring the commitments that govern the business process. The key enhancement here to traditional business process management is the introduction of the customer-supplier dialogue into business process design with particular emphasis on the integration of commitment management and related communications into the overall business process.

Part III

This part introduces process-centered management as the preferred approach for systematic, effective organizational response to planned or unexpected change. The concluding chapters of Part III are concerned with the implementation and deployment of highly adaptable business processes within a process-centered organizational framework. We try to give the reader the methods for understanding, planning, and implementing the process-centered organization, which is arguably the most effective construct for recognizing, handling, and absorbing change.

Appendix 1 presents some of the key tools for designing adaptable business processes and for implementing them in an integrated, timely, and cost-efficient manner. The Glossary, Endnotes, Bibliography, and Index complete the book, providing a set of references and source information to the interested reader.

Contents

Part I: The Case for Process Centering

Part II: Process Centering Fundamentals

Appendixes

Dedication

To Christine and Laura

Part I: The Case for Process Centering

1 Doing Business in the Face of Change

1.1 The Changing Business Environment

Doing business means providing products or services (or both) to customers who need them and are willing to pay for them based upon perceived value. The products and services are the result of a variety of organized processes — known as business processes — which represent all the communications and work carried out by the enterprise in order to produce and deliver. Within the business processes, work is always performed by resource-consuming entities, including people — whether it is in a manufacturing plant, engineering laboratory, law firm, hospital, university, government agency, restaurant, or a gas station.

1.1.1 The Fundamental Success Factors

Over time, changes in the way of doing business have occurred primarily in:

- Customer needs and expectations (the market)
- Suppliers' means of producing and delivering their products and services (suppliers' business processes)
- The relationship between the customer and supplier

The prime indicator of any organization's success in serving its constituencies — be it a for-profit enterprise or a non-profit charity — has always been *customer satisfaction.* In addition to customer satisfaction, success in the business world and the long-term viability of any enterprise depend on three fundamental success factors:

- Profitability
- Competitiveness
- Sustainable growth in created value

The three go together; while a profitable and competitive company may survive and stay in business for a while, success over an extended period of time is always identified with growth in value. This is key — while there are many ways an organization can grow, we will show later that *only growth in created value* can ensure continuing success.

1.1.2 The New Challenge: Accelerated and Unpredictable Change

In a static world, profitability and competitiveness will predictably lead to growth. However, the real business world is dynamic; it is characterized by a variety of changes which are increasingly unpredictable. The key challenge in today's world is the combination of the unpredictability of change and the accelerating rate at which it occurs. Change, unless dealt with quickly, will impact either the profitability or the competitiveness of the enterprise — or most probably both. As long as change is reasonably predictable, organizations can plan a variety of ways to deal with it. But when change becomes unpredictable, unplanned, or unscheduled, planned responses are no longer meaningful in light of the rapidity of change. The only effective response strategy is that of continual adaptation:

> *The measure of an enterprise's ability to adapt is a function of how quickly it can first sense and then interpret information available in the marketplace — to derive appropriate meaning from the torrent of noise it receives.*

This done, the information must be processed so the enterprise can decide what to do about the change represented by the information. This, in turn, means that humans deployed in the business processes of the enterprise must be able to make decisions about those processes — including decisions about the behavior and actions in the processes, the structure, and even the goals of those processes. Thus, they will create the capability for responding to the change sensed.

To respond successfully to change, one must:

- Recognize and understand change.
- Determine changes to current requirements and communicate them to those decision-makers who can make or change commitments.
- Transform processes to deliver the new products and services that will meet the new requirements in a satisfactory manner.
- Organize optimally the resources necessary to achieve this.

All this must be done rapidly so that the changes are implemented before another change occurs. By doing this, adapting to change becomes much easier. Dealing with one change at a time is always easier than making multiple changes simultaneously. It is also easier to pinpoint the causes of any failures when a particular change does not work as planned.

1.1.3 The Results of Change: What Is Really Happening?

Change is the natural state of the universe, only occasionally punctuated by brief periods of what we know as stability. Accordingly, the evolution of industry and commerce over the centuries has also been marked by ongoing, gradual, and mostly predictable change. In the last few decades, however, change has accelerated and become unpredictable to the point that it now presents a critical challenge to the managers of enterprises.

The goals, strategies, organizational concepts, and management methods of the traditional corporation have been made obsolete by the following:

- Virtually unrestricted and global flow of capital and, in particular, the unprecedented scale of capital flows and their ability to disrupt entire national economies[1]
- Emergence of regional trading blocks — such as the North American Free Trade Area (NAFTA), the European Union (EU), and Mercosur, the trading association of South American countries — with the progressive disappearance of boundaries between national markets
- Global deregulation of telecommunications, air transportation, and banking
- Unprecedented advances in information technology and the explosive growth of recorded, symbolic representation of information and knowledge such as the Internet — an enormous and pervasive conduit of ideas, information, and point-to-point worldwide contacts
- Emergence and phenomenal growth of knowledge companies conducting electronic commerce with intellectual, digitalized assets but practically no physical asset ownership
- Rapidly changing behavior of customers who now have more choices, and higher expectations, than ever before

Twentieth-century enterprises were organized around concepts that were superbly suited to the conditions of an earlier era; however, they are no longer able to perform well or to adapt and respond to change in the business environment of the 21st century. As a result, many of these same companies will be forced out of business, forced to merge to survive, or bought out because the world in which they have operated has changed beyond the limits of their capability to adapt or evolve. Restoring an enterprise's competitive position is no longer an issue of getting people to work harder or to produce better quality products; it involves learning why, when, and how to change the way work itself is done (its business processes). It is a matter of understanding that the processes of the organization are critical to its future. But how did we get to this point?

1.1.4 The History of Change

Today's companies are the products of change — a change that was much more predictable and more consistent in its evolution than what they are facing today. Until the industrial revolution, business between customer and supplier was conducted on a direct basis. Artisans — cobblers, tailors, blacksmiths, and the like — communicated individually with each of their customers to understand their needs and to discuss how the work would be done, what was to be delivered — and for how much. Goods and services were tailored to the unique needs and expectations of each individual customer, as every product was handcrafted. The relationship between customer and supplier was direct and personal.

1.1.4.1 Adam Smith and the Mechanization of Labor

The first major change in the evolution of the modern corporation occurred when Adam Smith published his historical book, *The Wealth of Nations,* in 1776. This work was based on the recognition that technology newly created by the industrial revolution (e.g., the steam engine and the factories powered by it) also created unprecedented opportunities to increase worker productivity in the manufacturing industries. One quantum leap in worker productivity was achieved through the consistent application of the principle of the division of labor. Today's corporations — petroleum companies, computer manufacturers, airlines, banks, accounting firms, steel makers, insurance companies, and hospitals — have all been organized around Adam Smith's central concept, the division or specialization of labor. The larger the organization, the more specialized the work assignments, resulting in fragmentation of the work. Specialists never complete an entire job or transaction; they just carry out piecemeal tasks or procedures. People working on an assembly line might become experts at installing doors on light-duty trucks, but they are unable to assemble the entire truck.

Specialization of labor, nevertheless, led to incredible increases in productivity through the mechanization of individual tasks. Its consequence was an impressive economic growth over the past 150 years. This growth occurred not only because of the dramatic increase in industrial productivity but also because of major changes created in the ways in which goods could be transported over land — complementing and enhancing the age-old maritime trade over water.

With the industrial revolution, the customer-supplier relationship also changed. The personal relationship between the customer and the artisan changed into a fragmented series of interactions between the customer and "the business." Increasingly, the customer had to deal with salesmen, service, and accounting people, rather than those producing the product.

1.1.4.2 Management System Based on Command and Control

In the 1820s, Europeans and Americans began building railroads, which first extended and then accelerated economic development through better distribution of goods and services. Railroad companies also made a major contribution to the technology of management. Borrowing a few basic principles from the military, they invented industrial operating procedures and the organization required to control them — known as the command and control system of management. This system grew out of the necessity to move traffic according to a pre-established, precise schedule and thereby prevent train collisions on heavily traveled, often single-track rail lines. Railway workers were drilled with military precision to act in accordance with the resulting work rules. The organizational structure required to create and then implement these rules became the hierarchical management structure of railroad companies; most hierarchical management systems in place today are still based on the concepts and principles introduced by the railroads over a century ago.

The industrial revolution brought with it the conceptualization of the enterprise as a machine. This *mechanistic* view, most recently described and analyzed by Ackoff,[2] suggests that enterprises, like machines, had no purpose or choice of their own; they had to perform the function of serving their owners' purposes. (Figure 1.1 shows a comparison of the four views of the enterprise as a system. Each view is characterized by a unique combination of the purpose and choice available to the enterprise as a whole, and to its parts by themselves.) The owners' principal goal, of course, was to earn maximum return on their investment which, in turn, meant that enterprises were expected to make a profit. Workers were employed only as long as they were capable of performing their assigned tasks. When they were no longer able to perform their tasks due to illness, injury, old age, or poor work habits, like parts in a machine, workers were replaced by younger, healthier, or more capable individuals. This was classical capitalism, pure and simple.

View	Purpose and Choice	
	Whole	Parts
Mechanistic	-	-
Organismic	✓	✓
Social	✓	✓
Ecological	-	✓

Figure 1.1. Independence of Purpose and Availability of Choice (Adapted from Ackoff, R.L., *The Democratic Corporation,* Oxford University Press, New York, 1993, p. 5. With permission.)

1.1.4.3 Contributions of the Auto Industry: The Assembly Line and Mass Production

The next significant evolutionary change in the development of today's corporation came from the automobile industry at the beginning of the 20th century. Henry Ford applied Adam Smith's concept to the production process by reducing each worker's job to a single task (such as machining a part or inserting and tightening a bolt) to be repeated in a precisely prescribed manner.

To further reduce the unproductive time spent by the workers in walking from one assembly to the next, Ford introduced the moving assembly line, which brought the work to the individual worker.[3] Breaking down the assembly process into a series of uncomplicated work steps made the job of each individual worker much simpler; however, at the same time the number of jobs increased by an order of magnitude. This made the process of coordinating the workers and of assembling the outcomes of their individual efforts into a whole product significantly more complex. Nevertheless, this approach worked reasonably well as long as the automaker was producing only one line of cars, such as Ford's Model T.

Coordination for a mix of cars and models, however, soon became an overwhelming task — in particular for General Motors, which began to produce a far greater variety of cars and models. When Alfred Sloan took over the management of General Motors, he created the corporate management system that was needed to apply Ford's more efficient production system to multiple lines of models and products. By establishing smaller, decentralized divisions for producing each car model and for making key components such as engines, generators, transmissions, and steering assemblies, Sloan essentially introduced the management system for mass production. It has existed ever since. Alfred Sloan applied Adam Smith's principle of the division of labor to the management process, just as Henry Ford had applied it earlier to the production process. The corporate executives and their new staffs of financial and marketing specialists complemented the company's engineers and manufacturing professionals, resulting in functional differentiation across the company.

1.1.4.4 Financing the Enterprise Through Public Ownership

Financing the expanding operations of the new industrial enterprises and the new technology of production required steadily increasing amounts of investment in plant and equipment. In the early days of industrial expansion, capital was not as accessible and borrowing not as attractive as today. Therefore, to leverage growth and productivity, many owners had to turn to the investing public for funds and raise capital by selling stock. With enterprises going public, ownership was transferred to the stockholders. This transfer handed control over the enterprise to a newly emerging managerial class. Making a profit came to be viewed as a means for fueling growth and ensuring survival, but was no longer the sole reason for the enterprise's existence.

Publicly owned enterprises became known as corporations, now with a purpose of their own: to survive and to grow. This world view then brought about a transition from the mechanistic to what is known as the *organismic* view[4] of the enterprise (Figure 1.1) which is still the dominant view today.

1.1.4.5 The Birth of Consumerism and the Modern Corporation

The functionally organized corporation faced its greatest challenge, as well as opportunity, between the two world wars and after World War II when the economic environment in the industrialized world was characterized by growing demand for domestic and foreign goods and services. Deprived of material goods, first by worldwide depression and then by World War II, the public worldwide was more than happy to buy whatever corporations were able to offer. And these corporations met the challenge beyond anybody's expectation — ushering in the era of *consumerism* and an unprecedented material standard of living in the 1950s and 1960s.

This enormous economic expansion, in turn, led to the next evolutionary change in the development of the modern corporation. In order to satisfy the relentless growth in market demand, elaborate planning models were introduced. Large staffs of corporate planners, controllers, and auditors were deployed. They were to determine which businesses the corporation wanted to be in, how much to invest in each, and how to optimize returns to be delivered to the corporation. This *make-and-sell* organizational model spread rapidly into Europe and then to Japan after World War II. Designed for heavy and growing international demand and, therefore, accelerating business growth, this form of corporate organization became typical for the emerging multinational corporations of the post-war era. They operated in an environment of steady growth, predictable consumer demand, and a supplier-driven market economy.

In the 1950s and 1960s, the main concern of company executives was capacity — that is, being able to satisfy the increasing demand. And this demand was primarily for consumable goods, rarely for quality or for attendant service. Consumers wanted things they did not have before: a new house, a new car, a new dishwasher, or a new television set — any make, any model was better than no model at all.

These new functionally organized companies were also well suited for effective managerial control. It was relatively easy to develop and approve operating plans and budgets function by function, department by department. Having broken work down into simple tasks, managers and supervisors could easily monitor and control work execution and thereby assure consistent and accurate performance against those plans.

Planning and control systems were developed to regulate product quality and to optimize delivery schedules and costs. The input-process-output (IPO) model and supporting techniques, such as statistical process control, were applied with increasing frequency to analyze and manage material and production processes.

Finally, as new computer technology became increasingly available in the 1960s and 1970s, corporations were able to fragment even more of their professional (e.g., white-collar) work into small, repeatable *procedures,* which could then be automated to a very large extent. Mainframe computers with terminals replaced forms, documents, worksheets, correspondence, and other paper-based representations of information and communications. Personal computers replaced mainframes. Voice-mail, pagers, and 800-numbers and automated answering devices often replaced person-to-person communication over the telephone. With these technological developments, more and more person-to-person contact within and outside the corporation was lost to online systems and automated applications such as order entry, billing, and reservation systems.

The net results of all these changes in a functionally organized corporation were

- The evolution of management systems containing organizational and communications barriers which impede the natural flow of work
- The complete disappearance of direct dialogue between customers and suppliers of products and services

The unprecedented economic expansion of the 1960s was paralleling significant social change and a shift in attitudes about human rights and workers' interests. People's views of work itself changed from its being a physical effort to a more knowledge-based and knowledge-dependent activity. As a consequence, corporate managers became aware of the need, in the interest of good labor relations and corporate performance, for taking into account the concerns, interests, and aspirations of their employees in the way they managed their enterprises. The concern with matching the purpose of the corporation with those of its employees gave rise to the *social view* of the enterprise, which attributes individual and collective roles and objectives to the people employed by it (Figure 1.1). This view holds that the corporation, as a *whole,* has a unique purpose and has a unique set of choices in pursuing it. At the same time, however, the *parts* of the corporation, including the people in it, also have their own and independent purposes, along with a set of choices available to them.

1.2 Where Are We Today?

1.2.1 Problems with Today's Corporation

Most of today's functionally organized, hierarchically managed organizations still are, virtually without exception, top heavy, resistant to change, and paralyzed by the inertia of middle management with vested interests in the *status quo* and outdated views on productivity.

1.2.1.1 Problem One: The Top-Heavy Organization

One problem is that the growth of organizational resources has not been proportional to the growth of demand: adding workers at the bottom of the fragmented organization also required the addition of a disproportionate number of managers, supervisors, and other coordinating staff. Fragmented organizations thus display a strikingly distorted resource distribution. The following example, first used by Hammer,[5] illustrates the point.

> *Let's assume that a manufacturing company does 100 units of work an hour, and each of its workers is capable of doing ten units an hour. Thus, the company would need 11 people: ten workers and one supervisor. Let's now assume that demand for the company's product grows tenfold to 1000 units of work an hour. In this case, the company would not need just ten times the number of workers plus one supervisor for each ten new workers. It would also need additional managers to manage the supervisors, as well as administrators and other staff to coordinate and expedite the work of the 100 production workers. Ultimately, the company would need close to 200 people: 100 workers, ten supervisors, one manager, and three assistant managers to manage approximately 80 people in a staff which includes such support functions as finance, audit and control, planning, human resources, facilitation, and expediting. The end result is an almost 20-times increase in resources to handle a tenfold increase in production.*

1.2.1.2 Problem Two: The Change-Resistant Organization

Another problem has been that traditional business organizations which fragment work processes tend to perpetuate the *status quo* and eventually become change-insensitive and even change-resistant. They stifle innovation and creativity because their functional structure was designed on the assumption that conditions will vary only within relatively narrow, predictable limits. As a result, there is no institutional recognition of significant change, and if any one particular manager or employee should happen to recognize it, the command and control mentality ensures that he or she will not be able to do too much about it.

1.2.1.3 Problem Three: Middle Management Inertia

Adding an increasingly large number of professional tasks to the already existing volume of manufacturing and production work, however, led to increasingly complex, difficult-to-manage processes of production or service delivery. The price companies had to pay for work simplification through fragmentation was the enormous growth of people resources in the middle of the corporate hierarchy: the middle management and corporate staffs, whose primary responsibility was to make an increasingly complex, hierarchical organization work. And this lopsided expansion was not easily reversible. When a company needed to contract, as corporations found out in the 1970s and 1980s, a proportional and rational reduction in labor and managerial resources was exceedingly difficult to accomplish.

1.2.1.4 Problem Four: Outdated View on Productivity

Until recently, economic theory has tied the concept of productivity and its related measurements to the production of mass and energy-based products and goods. In essence, productivity has been an industrial concept. As a consequence, traditional business management has taken a similar approach — with far-reaching consequences when it came to goal setting and the managerial actions to boost industrial activity.

Economists and managers have had a lot of difficulty in accounting for productivity growth in such service industries as banking, insurance, consulting, or legal services. The same goes for the creation of information and knowledge — the products of knowledge workers and knowledge organizations. The value created by either information or knowledge is intangible in terms of the classic productivity measurement of output per unit of labor or output per unit of time.

For decades, even as computers transformed work practices in almost every industry and helped create entire new categories of businesses and jobs, "measured productivity" in our economy — the output of goods and services per worker — barely advanced by a percentage point per year. This is because economic theory, while perfectly capable of measuring and accounting for the creation of capital assets, has been unable to develop consistent and credible measurements for the creation of such intellectual assets as information and knowledge.

Yet, more and more economic activity — the exchange of goods, services, and other assets for money — is generated by the "information economy," which has to be credited with not only value creation but also being the fastest growing segment of the national economy. It includes the myriad of services and products in banking, finance, and information services, as well as in the legal and consulting professions. Their products are ideas, decisions, solutions, creation or transfer of knowledge, and other outcomes of an informational or intellectual nature.

Federal Reserve chairman Alan Greenspan, along with noted economists, seems to be convinced that a fundamental change is under way in the economy. In his remarks to Congress in 1999, he stated clearly that the U.S. economy was enjoying "higher, technology-driven productivity growth." Several economists now seem to believe that current productivity growth derives from two sources:[6]

1. The economic value of quality improvement, speed, direct customer service, and the accelerated introduction of new and innovative products
2. New types of economic activity generated by the "information economy" — such as software development, consulting, technical support, electronic commerce, and training related to computer and communication technology

Added to this is the multiplying effect of network linkage among traditional back office processes — such as procurement, production planning, and distribution. These, then, seem to be the competitive advantages of information technology and the knowledge companies engaged in it. A few economists now feel that this new economy needs a broader definition of output and productivity — which goes beyond the industrial-era concept of parts and products coming off the assembly line.

1.2.2 Today's Challenge

These, then, are the principles and managerial approaches that have shaped today's corporations and the way in which they have conducted themselves. In the 1980s and early 1990s, companies faced very different and not necessarily evolutionary sets of changes. Now it is no longer sufficient for companies to organize their work around functional differentiation, managing it through an organizational hierarchy. Instead, companies must:

- Organize their work around the work itself, around their business processes.
- Design these business processes not only to be effective and efficient in delivering products and services in current terms, but also, and most importantly, to be adaptive to change — be it expected or unexpected.

Every organization has a payroll process. In most cases, it is a highly effective business process — incorrectly computed tax withholdings, missed pay days, or deposits to the wrong accounts are extremely rare. At the same time, today's payroll processes are highly cost efficient. With electronic funds transfer, they require a small fraction of the time that used to be necessary to print and distribute payroll checks. Most of them can be modified or updated online, resulting in considerable adaptability to changes in payee status or the tax law.

Today's highly competitive business environment is placing similar demands on business management. Customers continue to value cost and quality but consider them given. They demand more and more responsiveness, in terms of goods and services that satisfy their unique needs, delivered how and when they need them. Consumers have moved from requiring reliable, mass-produced products to expecting satisfaction from individually customized goods and services. Therefore, customer satisfaction must be the overriding objective if an enterprise wants to be competitive and successful.

> *The arrival of networked access to a worldwide set of suppliers revives and practically mandates an individual relationship between customer and supplier — not seen since the days of the craftsman and artisan.*

It seems, then, that an enterprise has to adopt the Mars Pathfinder team's approach: "better, faster, cheaper" in the face of any new set of requirements. The "better" refers not only to increased productivity and constantly improving quality, but also to direct customer-supplier interaction. "Faster" involves not only reducing cycle time, but also, and more importantly, being able to react and adapt quickly. And "cheaper" means reducing costs while increasing value. It is a tall order, but post-Challenger disaster NASA, using its reengineered processes, has lessons to teach us all: by discarding old paradigms and aggressively redoing processes, dramatic changes can be successfully managed.

1.3 Summary

Success in the business world depends on three fundamental factors: profitability, competitiveness, and sustainable growth in created value. The functional organization found in most corporations today is based on the concept of division of labor that was designed to operate in a stable, supplier-dominated business environment where change was continuous and highly predictable.

The key challenge in today's world is steadily accelerating, unpredictable change. Old-time, functional organizations have become inward-oriented, inefficient, and progressively ineffective in coping with the ongoing change in customer requirements — in many cases resulting in loss of market share and competitive position. Managed by the command and control concept, functionally organized enterprises are certainly in no position to respond to the onslaught of change — especially when that change is unpredictable.

The measure of the ability of an enterprise to adapt is a function of how quickly it can sense and interpret information available in the marketplace. It is no longer sufficient for companies to organize their work around functional differentiation and to manage it through an organization hierarchy. Instead, companies must organize their work around their business processes and design those business processes to be effective, efficient, and adaptive to change. The arrival of networked access to a worldwide set of suppliers revives the individual relationship between customer and supplier that is reminiscent of the days of the craftsman and artisan.

2 Traditional Ways of Coping with Change

2.1 The Nature and Sources of Change

The prime force driving today's companies deeper and faster into risky territory is relentless, accelerating, and unexpected change. Change, in turn, results from an environment whose complexity increases dramatically with the number of unique entities — individuals, institutions, enterprises, governments — directly interacting with each other. The accelerating rate of change results from the growing number and frequency of interactions such that change becomes virtually impossible to predict. An environment that is experiencing a simultaneous acceleration of unpredictable change and increasing complexity is said to be *turbulent*, bordering on the chaotic. Many organizations describe their operating environment as chaotic as business approaches the 21st century. Adapting a stable organization to change is the traditional goal of every executive and manager.

2.1.1 The Ping-Pong Response of Organizations to Change

All too often, organizations have reacted to external pressures and unexpected change with somewhat arbitrarily chosen, spasmodic, campaign-like initiatives to change such as:

- Automating existing operations
- Conducting company-wide quality awareness or customer sensitivity programs
- Training employees in statistical process control
- Training executives and managers in leadership — but without providing the means to transform the results of good leadership into good management
- Developing and communicating a business vision and mission — but without the means to translate them into sound plans and operations

- Embarking on culture change initiatives
- Reducing resources, outright downsizing, and other forms of job and cost elimination

These types of initiatives usually turn out to be temporary fixes and, at most, only partially successful because no one bothers to analyze what will happen to all the processes within the organization as each effort is launched. Like a well hit ping-pong ball that bounces erratically in every direction, organizations often suffer the same erratic and unpredictable results. Systems software exists to map the dynamics and impact of each new initiative. However, very few organizations are aware of its existence, so few use it or take the time to figure out what is happening to everything else in the organization as each new, seemingly simple and positive change is initiated. And so the effects of each change reverberate for months, even years, and no one suspects or understands what originally caused them.

Without addressing the fundamental factors that represent change and without the ability to activate the key business processes to accommodate and respond in a rapid manner, change will, at best, be temporarily deflected. In the final analysis, out of this turbulent or chaotic environment, change flows through two major, clearly identifiable channels: customers and competition. In each case, change is characterized by unpredictability.

2.1.2 Today's Customer-Driven Environment

Since the late 1970s in the developed countries of the world, the dominant force in the supplier-customer relationship has changed dramatically. The reasons for this change include:

- Customer experience and sophistication
- Customer choice resulting from competition
- Customer consolidation
- Customer proliferation

With increasing material wealth, customers now possess a much greater variety of material goods than half a century ago, in the heyday of supplier-driven, growing-demand economies. Now consumers own many things and tend to replace existing products and services with better ones rather than repair or renew those they had before. In doing so, they always demand quality: not only freedom from deficiencies (namely, that things work as they are supposed to, such as cars achieving the specified fuel economies), but also, and more importantly, meeting their unique and particular needs. Customers demand satisfaction features. These would include products that are tailored to their expectations, such as a vehicle specially equipped with a night-vision device, deliveries that match their work schedules, same-day delivery by appliance stores, or convenient payment terms such as "90 days same as cash." Market research shows that for most consumers, paying extra is not an issue if it saves them time (another satisfaction feature). For everyone in this day and age, time is the new natural resource of the 21st century, and people will do what is necessary to make sure that it is not wasted.

Saving time, and doing things faster, is the driver behind the many organized efforts to reduce the time it takes to complete a transaction, to produce a car, or deliver mail — efforts known as cycle time reduction. A textbook example of this is Motorola, where the manufacturing cycle time for pagers used to be 8 weeks. Over several years and with ongoing process improvement, it was eventually driven to below 2 hours — and that for the customer's model of choice!

2.1.2.1 Customization and Its Impact

In any market, consumer expectations soar when multiple competitors enter with new and unexpected products that are sold at competitive prices combined with higher quality. Under such circumstances, consumers will expect more and demand more because they know they can get more. Furthermore, the demands will become more frequent because the leading competitors have figured out how to introduce new products at a faster pace — namely, by reducing their own product development and production cycle time. Last, but not least, competitive success is increasingly dependent not so much on product quality or price, but on the levels of related services, distribution, delivery, payment terms and conditions, customer information and training, customer service, maintenance and repair — and, most recently, mass customization.

Today's fast-changing business environment mandates a profound change in the way products are defined. A new product can no longer be a fixed set of features to be sold to a mass of customers without variation. The demand in the future will not be for products but for sets of features custom assembled to meet unique characteristics of individual buyers so as to create maximum value in each and every case.

People can now order clothes or shoes made to fit their exact physical dimensions. Levi Strauss has created a customized order process to provide tailored-to-fit jeans. Custom Foot makes a pair of custom-fitted shoes at a competitive price. The shoes are produced on the basis of a digital model of *each* of the customer's feet.

DayTimers, the Allentown, PA, supplier of calendars, organizers, and professional planners, now allows its customers to personalize its standard planners. Customers can identify the dates of important events when they place their order, and these dates are then preprinted on the corresponding calendar pages. At a negligible additional service cost, delivery time is the same as for the standard version. Futurists tell us most products will not be produced until there is a specific need for them. We will have only two markets: custom made and used.

2.1.2.2 New Businesses Generated by Shifting Consumer Needs

There has been an impressive consolidation of institutional customers in certain markets. The United States has experienced the emergence of:

- Retirement funds in the securities industry
- Mega-dealers in the automobile business, such as Family Auto
- Fast-food franchises such as McDonald's replacing thousands of independent cafes and restaurants
- Warehouse-sized discounters such as Wal-Mart that have driven family stores and small retailers out of business

All of these changes have also dramatically impacted the customer-supplier relationship.

John Smith's computer superstore offers IBM, Hewlett-Packard, Apple, Compaq, and Dell products. Because John Smith offers more than just IBM computers, he is in control of his relationship with IBM. With so many competitive sources available to him, John Smith needs IBM, but has the upper hand in the relationship because, if IBM left, he still has some very competitive machines to sell. In his negotiations with IBM, he can more readily force IBM to meet his changing requirements for functionality, quality, commercial terms and conditions, and price. If he sold only IBM computers, his negotiating advantage would disappear, as IBM would try to dictate more terms because John's store was a franchise store.

Companies have increasingly recognized that it is in their best interest to satisfy the needs and expectations of a growing number of stakeholders. Accordingly, these stakeholders are considered as customers: basically all those who are affected by the company's products, services, or processes — employees, suppliers, investors and creditors, regulators, government, the media, and even the general public. As a result, the number of those who need to be treated as customers has proliferated because businesses have recognized their importance in their long-term success.

2.1.3 Competition in the 21st Century

The other source of change, competition, is also very different from what it was even 20 or 30 years ago. This difference can be derived from these factors:

- The competitors
- Information technology
- Nature of competition itself
- Suboptimization

2.1.3.1 The Competitors

With the globalization of the economy, the pace of change has accelerated, because individual companies face a greater number of competitors, each of which is capable of introducing more products and services to the market. This acceleration is resulting in a significantly larger volume of competitive challenges to respond to. The volume of competitive change, in turn, promotes more frequent responses and requires faster innovation and significantly shorter development and production cycles. As a result, product life cycles with products such as upgrades of computer software programs have gone from years to mere months or sometimes even less.

Not only have product and service life cycles diminished, but so, too, has the time available for innovation, product development, production, and marketing. Today, competitiveness means that a company must recognize change almost instantaneously and then respond quickly or it will not be able to respond at all. Most importantly, it also has to anticipate and recognize change far beyond that which is expected. The changes that will create a crisis and may eventually put a company out of business are those that happen outside the realm of its current expectations — the type of unpredictable change that dominates today's business environment.

2.1.3.2 Information Technology

Technology has also changed the nature of competition in unexpected ways. In retailing, it allows manufacturers and retailers to merge their distribution and inventory systems in mutually beneficial ways. Federal Express Company, for example, has developed an ingenious information system for managing the enormous task of accepting, registering, tracking, and delivering millions of letters, packets, and packages every day around the world. Customer service representatives, whether by telephone or in person, can address an inquiry already briefed on the nature of the problem and the particular item's entire history by consulting an ever-present, continuous, precise, updated order process, as well as an inventory management and distribution system. This approach allows Federal Express, above all its competitors, to not only satisfy virtually every whim and need of its customers but also raise customer expectations for all of the delivery companies in the worldwide market.

Beyond the integration of intra-enterprise systems and the online buying and selling of goods and services over the Internet, known as *e*-commerce, however, information technology has brought about something much more important. It is fundamentally transforming the way an enterprise does business by creating a fully integrated value chain that comprises the enterprise, its suppliers, and its customers. This approach to doing business, known as *e*-business, goes deep into the business processes and culture of all its participants in order to optimize all activities involved in delivering greater perceived value to the ultimate user. Those companies that do not master the Internet's universal connectivity and do not create an integrated value chain will not be expected to stay competitive.

2.1.3.3 The Nature of Competition

The nature of competition has been changed by, among other factors, the appearance of increasing numbers of start-up players in every major market. But start-ups usually do not do business in the traditional way and do not pattern themselves after traditional companies; they create new rules and fashion themselves a new image. Just as Wal-Mart did not create itself in the image of Sears, Microsoft did not follow in IBM's footsteps.

2.1.3.4 Suboptimization: Impediment to Competitiveness

Several examples show that the partial results organizations achieve are often very different from the overall results that their management desires. Often the efficiency of one particular operation comes at the expense of the efficiency of the whole.

A major computer manufacturer set a goal of filling special customer orders ahead of competition — but this goal was proving elusive. Like many companies in its industry, the company used a multifunctional special-order approval process. But, while the order approval process was improved dramatically (reducing the average cycle time from 14 weeks to 17 days), the product development and production processes (which reported to a different management hierarchy) deteriorated. This happened because they were slowly overwhelmed by the increasing number of special orders flowing from the reengineered order approval process. The computer manufacturer's approach is clearly an example of process suboptimization.

In many cases, rigorously observing company policy and managerial guidelines comes at the expense of customer satisfaction and even the loss of revenue, as in this real-life example from air transportation:

> *A plane belonging to a major North American airline developed mechanical problems en route and therefore was unable to proceed to its immediate destination in Atlanta, where it would have taken on passengers before proceeding to its final destination in Sarasota. Passengers awaiting the plane in Atlanta were told that there would be some delay in their departure for Sarasota. The airline had an airplane available at the Atlanta airport, but without a crew. The crew to fly to Sarasota was on the delayed plane that had developed the mechanical trouble. Several flight attendants based in Atlanta could have been called to form a new crew for the Sarasota flight. Although all of these flight attendants had just come off flights and, therefore, were entitled to free time, some of them volunteered to take the available plane from Atlanta to Sarasota. One attendant, however, refused, citing labor rules and her employment contract. As a result, an adequate crew could not be assembled and the available plane could not take the waiting passengers to their destination, Sarasota. Customers had to stay at local hotels and then were dispatched to Sarasota early the following morning, by which time a crew was available to fly an airplane there.*

This case is another example of suboptimization. While labor rules and the contract were scrupulously observed and ultimately satisfied, a significant revenue loss was incurred by the airline. Nor did the airline's management have any excess capacity for such contingencies built into their staffing plan. The situation was made worse by the dissatisfaction of many customers who could have taken flights offered by other airlines had they not been asked to wait for the resolution of this problem. Because of the inability to resolve the problem satisfactorily in a timely manner, a multi-million dollar aircraft sat idle, and the airline lost tens of thousands of dollars in revenue and incurred thousands of additional dollars in hotel expense. The manager who was ultimately responsible for this decision at the Atlanta airport was neither foolish nor unresponsive to customer requirements; he was simply doing what he was supposed to be doing — following company policy as it applied to labor rules and observing Federal Aviation Administration requirements on minimum crew size. Of course, adherence to the rules probably has lost a considerable number of customers long term.

These stories are not exceptions; they are often the rules. This is not how corporate executives want their organizations to behave, yet these follies can be repeated endlessly as this behavior persists all over. Why?

2.2 Traditional Responses to Change

In the 1970s, the emphasis in American industry was on productivity, and, indeed, some significant gains through massive investment in new technology and wage settlements more conducive to productivity were made. American managers soon realized, however, that our foreign competitors — especially the Japanese — were able to steadily match our improvements in productivity. Furthermore, while American management was betting heavily on productivity and efficiency, Japan continued to outperform us in the area of product and service quality.

To counter this challenge, American companies have undertaken a number of remarkably similar initiatives over the past decades. They launched total quality management (TQM) initiatives in the late 1970s and early forms of business process improvement in the 1980s. While many of these initiatives showed tangible results, they rarely led to an overall improvement of business performance. As competition increasingly became an issue of price and cost, organizations started business process reengineering. Others undertook "downsizing" and "rightsizing" as part of various cost-reduction programs. Concurrently, companies have been in an almost continuous state of reorganization in the midst of organizational effectiveness initiatives. But none of these initiatives was designed to help companies address and respond to accelerating and increasingly unpredictable change.

Of the more than 40 companies listed as models of excellence for our business age by Tom Peters and Robert Waterman in their landmark 1982 book, *In Search of Excellence*,[1] less than a dozen are still considered successful today. Some have disappeared or have been taken over, some have been disasters, and many more are struggling to just stay alive. Some of the casualties from *In Search of Excellence* were once technology leaders: Amdahl, National Semiconductor, Wang, and Digital Equipment. The bottom-line diagnosis for all of them: inability to adapt to accelerating change.

As a result, corporate management is finding itself in a perplexing dilemma. Companies are now realizing that being committed to continuous improvement at both the micro (e.g., problem-solving) and macro (e.g., reengineering) levels, as well as to cultural transformation (e.g., change management and organizational effectiveness) is not what will make them competitive in the future. The next crucial step in their search to be competitive must first answer the key question: "What next?"

2.2.1 Total Quality Management

Attention shifted to total quality management (TQM) in the late 1970s and early 1980s. TQM not only resulted in improvements in product and service quality, but also contributed to changing attitudes about quality — that is, the understanding of quality, as well as the mentality and management approach necessary to make it happen in a continuing and pervasive manner. Managers realized that quality is definable and measurable, that objectives and requirements can be formulated toward its achievement, and that specific projects can be implemented to ensure that quality is improved where that is a necessity. In short, change is manageable as long as it has a predictable rate and direction.

Total quality management has been based on experience and practical observation; it is an empirical management approach. As a result, TQM is not a formal system, but a structured aggregate of its key components:

- Fundamental concepts
- Infrastructure
- Managerial processes

By design, TQM establishes temporary structures (project teams) within a traditional organization that it assumes to have a stable structure. Accordingly, TQM does not provide any theory or methodology to guide managers in creating new organizational structures. The fact is, however, that the average elapsed time between reorganizations in most American companies is less than the time required for fully implementing TQM. Consequently, organizational or management change is frequently a major impediment to the completion of a TQM initiative.

Like any management methodology, TQM is most effective when used within its intended range of application. Traditional quality improvement is designed for the removal of deficiencies and waste. Accordingly, continuous improvement is intended to move its objects (products or services) incrementally, project by project, and in no other way[2] — from where they are to where one wants them to be. In situations where the deficiencies inhibit desirable properties or behavior, continuous improvement results in demonstrably better products and services. However, in a significant number of cases, defect removal in itself provides no assurance of achieving a higher level of customer satisfaction than what it was at the outset. This is because either the customers were initially dissatisfied or by then they may want something entirely different! The lesson learned from all this is that quality has two faces:

- Freedom from deficiencies
- Satisfaction features

To be successful, quality initiatives should primarily address what customers want. Implemented and managed on this basis, TQM with its consistent problem-solving and planning capabilities, as well as team-based culture, provides a disciplined and well-structured base upon which further management approaches and systems can be built.

Continuous quality improvement, as traditionally implemented in functionally organized companies, nevertheless has its natural limits due to two inherent reasons:

1. Philosophically speaking, continuous improvement is based on continuous rethinking: it involves relatively small changes made close together in time. This approach, however, does not deal effectively with accelerating and unpredictable change or situations calling for a major, creative response — such as entering a new market, redesigning an entire product line, or financing a major acquisition.
2. Enterprises tend not to take a broad enough view of continuous improvement. Deficiencies in different parts of the organization are usually observed independent of each other, then removed by means of separate projects. However, the overall performance of a system (which an enterprise is) depends on the interaction of its parts, not on how the parts act by themselves. In other words, the performance of the organization as a whole will not necessarily improve when the performance of its parts taken separately is improved. (Ignoring this concept leads to the process paradox described later in this chapter.)

Over the years, managers have learned that a series of incremental, but not always related improvement projects may reach the point of diminishing returns. What this means is that, beyond a certain point, the success of quality initiatives cannot be measured solely by the number of people participating, the number of quality management teams launched, or even by the aggregate sum of costs saved or avoided. The success of TQM initiatives in the final analysis must be measurable by enterprise-wide performance indicators, such as:

- Increased speed of bringing products and services to the market (reduced cycle time)
- Growing market share and customer base
- Profitability
- Enterprise adaptiveness: the organization's ability to respond to unexpected change more rapidly and effectively than competition

To make this happen, practitioners of TQM realized that they needed advanced methods and more powerful tools beyond their basic quality management toolkit. As a result, TQM has been progressively linked to strategic planning, cost management, empowerment, and organizational effectiveness; the TQM toolkit has been extended to include benchmarking, process analysis, customer loyalty assessment, and a variety of team-building and facilitation techniques. The most important extension of TQM, however, came through the seamless transition to business process management and, in particular, to business process reengineering.

2.2.2 Business Process Reengineering

The 1980s taught us that business success depended largely on providing satisfaction at the lowest cost. Quality, effectiveness, cost, and efficiency became the competitive battle cries; working faster was just as important and profitable as working better with less. This premise, in turn, gave rise to traditional business process management,

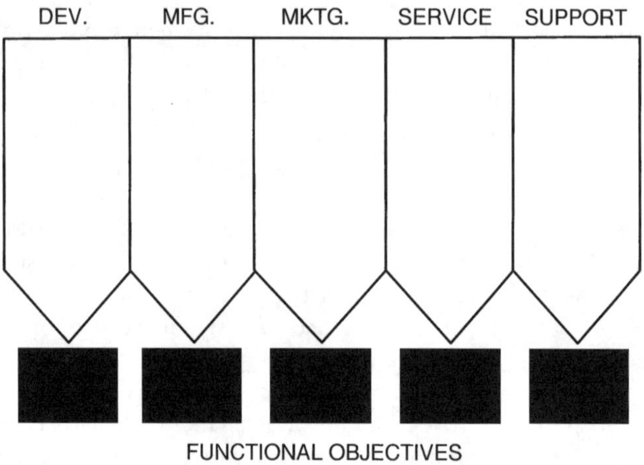

Figure 2.1. Functional Management System

gained a much better understanding of the idea of reengineering and its business importance to their companies.

Most of these companies have clearly delineated functions: vertical silos — structures built around the concept of the division of labor and hierarchical management. These functional silos comprise only narrow pieces of a multitude of processes that in totality make up the workings of the enterprise. Figure 2.1 is a schematic of the typical, functionally organized manufacturing company, with its five traditional functions shown as vertical silos operating in virtual separation from each other. Managers and executives are in charge of increasingly large chunks of the function, but none is in charge of the processes. This is known as process fragmentation. Because of process fragmentation, hand-off problems proliferate.

Typical hand-off situations occur when the work product of one function is handed over to the next function in line. For example, the sales function regularly produces sales forecasts. They are then presented to the manufacturing function which will plan production schedules against them. In a typical functionally organized company, the sales forecast is reviewed several times by the entire vertical hierarchy of sales management before it is released to the manufacturing people. These reviews, internal to the sales function, develop their own dynamic, their own rules and schedules — and as a result, may take excessively long time cycles to complete. In the meantime, manufacturing capacity is held idle and the production schedule may have to be delayed.

As a result, processes become highly error prone as each hand-off problem requires coordination and communication and entails wait times and mismatches in schedules. And, while these processes have always existed (they are, in essence, the way work is done by the enterprise), only a few people are aware of them, and they are certainly not managed in any consistent or rational way. To overcome the

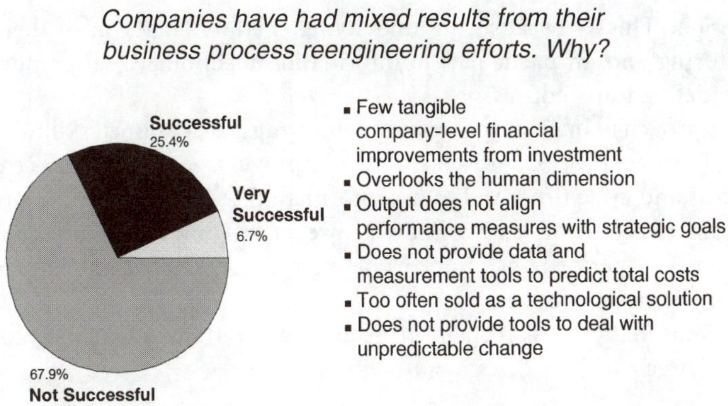

Figure 2.2. Business Process Reengineering

problems of fragmentation, business process reengineering establishes permanent structures — process ownership, process management teams, and the business processes themselves — in a traditional organization which it then intends to change.

Years of process reengineering have also brought the realization that as many as two-thirds of all reengineering efforts fail to reach their initial objectives — thereby further aggravating an already challenging situation (Figure 2.2).[4] The reasons for failure are many, but most can be avoided or corrected.

2.2.3 Limitations of Traditional Reengineering

Too many process redesign efforts ended with the question from the process team, process owner, and the stakeholders: "What can we do to ensure adaptability to future change, short of another major redesign?" The traditional process reengineering movement evolved around the concept of physical workflow and its improvement and redesign. This is because the underlying methodologies have their roots in manufacturing and, in particular, in information technology, whose purpose is to automate defined procedures.

This approach to business process reengineering has developed skillful ways of redesigning business processes to measurably improve product and service quality, to reduce costs and cycle time, and to use information technology for the streamlining and coordination of complex workflow. However, as companies embarked on using this approach to performance improvement, it eventually became clear that some things were being overlooked.

The traditional process reengineering approach did not adequately address several critical areas:

- The human dimension
- The flow of information and communications in the process

and redesign. This is because the underlying methodologies have their roots in manufacturing and, in particular, in information technology, whose purpose is to automate defined procedures.

This approach to business process reengineering has developed skillful ways of redesigning business processes to measurably improve product and service quality, to reduce costs and cycle time, and to use information technology for the streamlining and coordination of complex workflow. However, as companies embarked on using this approach to performance improvement, it eventually became clear that some things were being overlooked.

The traditional process reengineering approach did not adequately address several critical areas:

- The human dimension
- The flow of information and communications in the process
- The planned adaptability of redesigned processes to deal with unpredictable change
- Overall economic and business performance of the enterprise

What is needed is a very different management approach that will ensure the driving role of empowered, accountable people within individual business processes, and the systematic realignment of the enterprise around its key, adaptable processes.

2.2.3.1 Overlooking the Human Dimension

Overlooking the human dimension — mobilizing and motivating people — has been viewed as a major shortcoming of classic workflow-based process reengineering. The downsizing done without concern for the people has made the term reengineering imply that people do not matter because the approach taken is mechanical and impersonal. It is not surprising that, as a result, reengineering has been — erroneously and unfairly — identified with downsizing, layoffs, and job elimination.

2.2.3.2 Difficulty with the Communication Flow

Another limitation of traditional reengineering is its inherent difficulty dealing with the great variety of intellectual, managerial, and professional service processes where there are no easily definable physical work products (e.g., most work results are agreements, decisions, commitments, and the transfer of information and knowledge). In these processes, most of the work consists of a series of dialogues and related information exchanges.

2.2.3.3 Problem with Unpredictable Change

One of the greatest challenges to BPR is that it must support thinking that deals with the acceleration and unpredictability of change — the ability to identify and abandon old paradigms which underlie current operations. Most importantly, at the heart of this challenge lies the need to respond to change by redefining existing commitments based on assumptions about technology, people, and organizational goals that are no longer valid. Unless organizations can recreate these commitments or create new ones, any interim reorganization or process realignment they propose will be no more effective than rearranging the deck chairs on the sinking Titanic.

2.2.3.4 Uncertain Contribution to Overall Business Performance

The experience of the last 10 years clearly shows that reengineering of business processes for operational effectiveness and efficiency is a necessary, but not sufficient, condition for sustained competitive superiority. Nor is there any method currently in use which assures that reengineered processes not only will be effective and efficient in current terms but will also be highly adaptable and responsive to unexpected change. This is a clear limitation of most of the popular and broadly implemented process reengineering methodologies — be they embraced and promoted by individual gurus or the respected big five business consulting firms. Something else, something more, is still needed.

2.3 Limitations of the Traditional Approaches to Process Change

As we have seen, most of the influential business trends of the last three decades of the 20th century have focused on some form of process change. Quality improvement, process redesign, cycle time reduction, team-based organizational effectiveness, and other similar managerial approaches were essentially all process-related initiatives. There are several observations that will be useful at this point:

- Total quality management (TQM) speaks of repairing broken processes and emphasizes their continuous, incremental improvement.
- Process reengineering proposes to replace or radically change existing processes, broken or not.
- Cycle time reduction is essentially the streamlining of the work flow to do the same or less in a significantly shorter period of time.
- The team-based organization is designed to achieve more flexible, collaborative, and communicative work processes.

Each of these approaches shows how to get a particular aspect of a process right in a particular manner; most are based on the assumption that an enterprise will improve simply by refocusing all of its processes to their respective customers' needs. There are two points here, however, which deserve further examination:

1. Meeting customer requirements often is not enough; customers are not as good at telling us the specifics of what they want as we or they first thought.
2. Many companies that implemented business process reengineering and have shown significant operational improvements eventually went into economic decline and even bankruptcy — the process paradox.

2.3.1 The Imperfection of Customer Needs

While customer needs and requirements may be more predicable, they certainly do not represent the total picture. Customers often depend on their suppliers to help them stay ahead of their competition and to be on the cutting edge of innovation. Nobody said they needed a photocopier, a fax machine, a snowmobile, or a cellular telephone. No one needed the digital computer because no one knew that there were problems it could solve — until it, in fact, appeared. The source of less predictable change is often the competition or unexpected technological breakthroughs in the suppliers' processes.

IBM's troubles started when it initiated the personal computer revolution but failed to recognize the impact of user-friendly software on the marketplace — as the driver, not just a support component. One of the established telephone companies' greatest challenges arose when they missed the cellular telephone explosion and did not recognize the impact of radiotelephony on their land-based networks.

New technology usually creates a demand and market for new uses and applications that had previously not existed. In other words, an important new technology first uncovers the problem and then solves it. At that point, the latent need suddenly becomes obvious and urgent.

2.3.2 The Process Paradox

Even the seemingly best companies may have significant problems. It is now well known that a firm mentioned in Michael Hammer's article about business process reengineering[5] — an insurance company named Mutual Benefit — had gone out of business soon after the article was published. Other industrial corporations that embraced business process reengineering (or the same concept under a different name) have shown little or no overall performance improvement, even though individual process reengineering efforts were considered highly successful. These companies include IBM, Alcoa, American Cyanamid, and several others.

Success at the business process level, but no improvement or even negative overall impact at the enterprise level, is the essence of what has become known as the *process paradox.*[6] In other words, process improvement can lead to tangible local success; at the same time, it can also lead to significant disarray at the enterprise level. Interestingly enough, those who are responsible are often not directly implicated, so ultimate accountability for the ensuing problem is not clear, either. As we will see later, accountability is key to the success of process-centered management; therefore, the issue of who owns the final outcome must be carefully monitored and analyzed. If lessons have to be learned, this should be done correctly, clearly, and fairly.

After months or even years of careful process redesign, many companies have achieved impressive improvements in individual business processes — only to watch overall results decline. By now, paradoxical outcomes of this kind have become almost commonplace:

- IBM's Small Business Division won the Malcolm Baldrige Award in 1992 in recognition of the successful redesign of the development and production processes that produced one of IBM's all-time success products: the AS400. At the same time, IBM Corporation was facing the greatest crisis of its history, including the downsizing of half of its workforce worldwide.
- Another Baldrige winner, Florida Power & Light, was one of the textbook examples of total quality management, cutting the average annual per customer power outage from seven days to less than one day. While achieving this impressive result in its customer service, it created a new quality bureaucracy and increased internal costs.
- The Baldrige winner of 1994 in the small-manufacturing company category was forced to declare bankruptcy just one year after that.
- Management of a major international construction company proclaimed a 20% cost reduction and 25% quality improvement at the completion of a series of process improvement projects — yet, in the same period, overall business costs increased and profits declined.

These examples of the process paradox ought to be carefully studied by any executive who has initiated efforts to improve or reengineer individual business processes, only to discover later that overall business performance remains stagnant or, worse, even declines.

2.4 What Now? A Change in Managerial Attitudes

Effective response to change will require a change in *managerial attitudes* about change itself. Every management team will need:

- An *understanding of what change really is.* Management needs to realize that change is definable and, many times, measurable, that new requirements and objectives can be formulated as a result of it, and that specific work efforts can be executed to ensure that requirements are met — in short, that change is manageable.
- *A greater awareness of change.* Change is natural but inherently risky; nevertheless, it also has potential benefits. There must be a realization that in the final analysis, the ability to respond to change in an effective manner depends on adaptable business processes.

- The *realization that managing change will significantly contribute to competitiveness and, ultimately, to profits*. The erroneous perception is that change is a threat, or, at best, a necessary evil that must be coped with in some fashion. It must be replaced by the realization that every dollar invested in adaptability will improve the bottom line, as well as make the organization more competitive.

It has often been suggested that companies could be more responsive by changing their corporate strategies on an ongoing basis. For example, they should divest themselves of one business and acquire another; get into a new market and get out of another; change the nature, scope, or image of their business, and so on. But this kind of thinking distracts management from making fundamental changes in *how* they actually do the work — just as changing one's clothing does not necessarily change behavior and the things one does.

Companies are not just a collection of holdings or portfolios of financial assets, but also of resources. Among them are people working together through a variety of processes to create, make, sell, and distribute products and services. If they fail, it is because their processes cannot be adapted to create, sell, and/or distribute the new products and services called for by the change. Acquisitions or divestments will not materially change business processes and, in many cases, may even further aggravate already unresponsive performance.

Another suggestion is that companies would better respond to change if they eliminated management deficiencies, that they would succeed if their management system was different. But none of the fashionable management movements of the last 30 years — not management by objectives, management by contention, decentralization, matrix management, management by walking around, functional excellence, total quality management, or traditional process reengineering — has materially improved corporate ability to respond to unpredictable change and thereby improve competitive performance. Only the *planned adaptability* of a process will enable it to respond to unpredictable change — through the rapid but disciplined reconfiguration of existing and new commitments.

The key question, then, is: Does a company have the right products and services available for the right need at the right time and always ahead of the competition? The chapters which follow will address what has been learned about designing adaptable processes for effectively responding to change, including:

- Overcoming the common pitfalls in traditional process redesign
- Understanding key factors that determine process adaptability and, therefore, organizational responsiveness to change
- Integrating human resources and information technology into the response to change

2.5 Summary

Adapting a stable organization to change is the goal of every executive and manager. Because of the complexity of its origin, the rate of change is accelerating, and change itself is becoming increasingly unpredictable. Traditional improvement or reengineering methodologies, such as TQM, cycle time reduction, or process reengineering, will address the effectiveness and efficiency problem in current terms but will not result in new processes that are highly adaptive and capable of responding to unpredictable change. Companies are now realizing that being committed to continuous improvement at both the micro and macro levels, and to cultural transformation, is not what will make them competitive in the future. Success at the business process level, but no improvement or even negative overall impact at the enterprise level, is the essence of the process paradox. In other words, process improvement can lead to tangible success; at the same time, it can also lead to significant disarray at the enterprise level. What is needed is a very different management approach that will ensure the driving role of empowered, accountable people within individual business processes, and the systematic realignment of the enterprise around its key, adaptable processes. Only a complete transformation[7] of business processes, which are defined as the integration of commitment communications and workflow, can result in new designs that will be highly adaptive and responsive.

3 The Economic Value in Process Centering

In the face of new demands or unexpected change, companies have been inclined to launch massive, but mostly uncoordinated, multiple initiatives which then may or may not may not achieve the desired goal. Furthermore, these efforts can seriously impact the financial and human resources of the organization, especially if the change initiatives are strategically incompatible.

A medium-sized chemical company introduced a pervasive total quality management (TQM) initiative in its manufacturing operations and materials purchasing department. At the same time, an equally aggressive organizational effectiveness (OE) program was launched in its sales organization to improve teamwork and morale. Parallel, but unrelated to these two efforts, was the massive automation of the company's financial systems (billing, accounts payable, and so on). Each effort, separately, showed significant results. In the end, though, the lack of strategic coordination among the three efforts caused near-chaos, undoing some of the accomplishments. Eventually, these efforts were redone under the broader umbrella of process centering.

The process-centered response to change is to focus decisions, commitments, and resources on a small number of major opportunities by investing in only those processes that will make a considerable difference. The process investment problem, for which this chapter offers a conceptual guide for action, is how to identify and select processes to be reengineered and then leveraged through process-centered alignment.

3.1 Managing Work for Value Enhancement

Since the early days of economic development, the purpose of all forms of work organization (firms, corporations, processes) has been *value creation.*[1] The capability

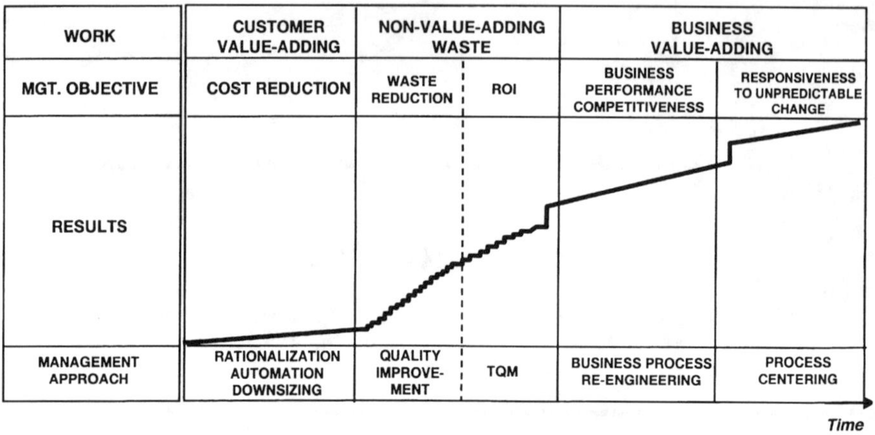

Figure 3.1. The History of Managing Work and Value

of producing value is also a prerequisite to growth — although not all growth means increase in value.

3.1.1 *The Value Contribution of Work*

Depending on how successful or unsuccessful an organization has been in attaining its prime purpose, its management will be motivated to find ways of enhancing or improving its basic value creation methods and mechanisms. The history of managing work over the past century shows a continuing search for increasingly effective methods for creating net added value (Figure 3.1). Until very recently, the framework in which work was managed has been provided by the functionally organized firm or corporation.

In terms of who is the beneficiary of the added value created, work itself can be conveniently divided into three categories:[2]

1. Customer value (CV)
2. Business (organization) value (BV)
3. No value (NV; waste)

Work activities that create added customer value are defined as those for which a customer (the recipient of the work product or service) is willing to pay compensation. This category of work is conducted primarily in the enterprise's core processes — such as product development, marketing, manufacturing, selling, distribution, or customer service (shaded area in Figure 3.2). Work that creates added business value is defined as that which is necessary for either the conduct of business, such as accounts payable, or the creation of assets for the business, such as capital investment planning. This type of work is normally carried out in the support and strategic (asset management) processes (Figure 3.2) of the enterprise.

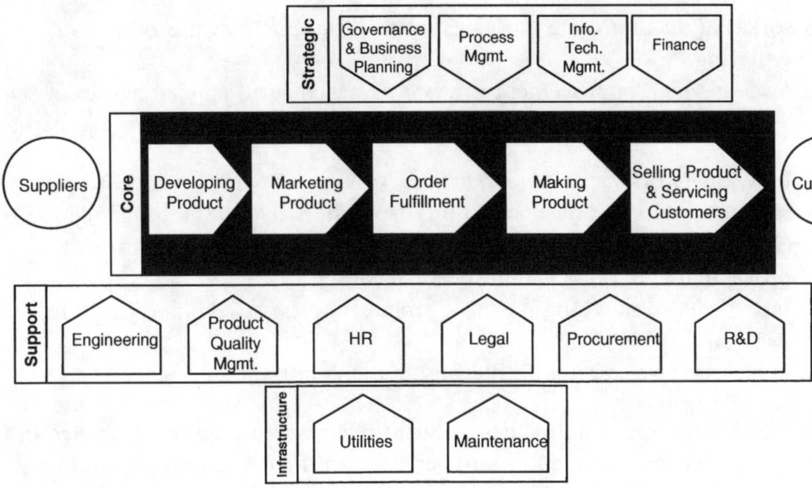

Figure 3.2. Creating Customer Value in the Process-Centered Enterprise

Work that creates neither customer nor business value (but may or may not have to be done for other reasons) is termed no value-added work, or simply waste. Waste may represent work intended to create customer value but which did not meet agreed-to conditions of satisfaction and was, therefore, found unacceptable by the customer.

Early work-enhancement efforts between the two world wars and in the 1950s and 1960s primarily targeted customer value-adding work (Figure 3.1). The objective of these efforts was more rational work organization (rationalization), improved quality and productivity (mechanization and automation), and, eventually, improved efficiency through cost and capacity reduction (downsizing).

Classic quality improvement and TQM aimed primarily at the removal of deficiencies and waste by reducing or eliminating, one by one, all identifiable no-value-adding work activities (Figure 3.1). Because of the relatively modest level of capital investment necessary to achieve tangible and often significant cost savings, TQM has also been found to be a very effective approach to achieving high rates of return on short-term investments.

Business process improvement and its more radical form, process reengineering, were initially intended to significantly improve, in current terms, the business performance and competitiveness of the enterprise. They achieve this by the systematic redesign or replacement of key asset-creating or support processes: the aggregations of business-value-adding work. The wholesale reengineering of planning, information systems, accounting, payroll, and procurement processes in the early 1990s at the beginning of the reengineering era is the well-known result of this approach. More recently, management discovered that business performance can further be improved by the reengineering of its core processes: the aggregations of customer-value-adding work. This focus on core processes is also beginning to lay the foundations for process centering.

The level at which improvement to work management is applied has also risen consistently within organizations. Over time, it has followed this evolving pattern:

1. Individual work steps and tasks at the very bottom of the hierarchy (Taylorism, rationalization, and mechanization), from the early 1900s through the 1950s
2. Organizational units, functions, and cross-functional projects (quality improvement and TQM), from the 1920s to the 1980s
3. Individual, overarching business processes (process reengineering), in the 1980s and 1990s
4. Enterprise level (process-centering), moving into the 21st century

One explanation for this evolution lies in the fact that, as we will see later, change and its effects have become more and more pervasive in the corporation as they affect successively higher levels of the hierarchical organization — until they finally impact it as a whole.

3.1.2 The Questionable Value Contribution of Downsizing

For many years, management theorists have been telling corporate executives to reduce capacity and to shrink their organizations as necessary measures in difficult economic times. The benefits of downsizing were extolled as the surest way of becoming leaner, more efficient, and profitable. Many companies in crisis did not really have any other choice, or chose not to look for any alternatives. Some considered continuous improvement too slow, while others felt process reengineering was too complex and risky. But in the final analysis, the overall result was mixed at best. Companies may have become leaner and smaller — but not necessarily fitter.

A survey by the American Management Association in the mid-1990s found that less than 45% of downsized companies reported any increase in operating profits. This means that downsizing did not bring increased cash flow or reserves, which would allow the company to begin making new investments toward a competitive recovery and growth. In plain English, it is questionable whether downsizing can be expected to add net economic value to the enterprise. The same survey also indicated that the attendant effects of downsizing included a decline in morale and the loss of the most productive and entrepreneurial employees — resulting in a further negative impact on productivity, profits, and intellectual capital.[3]

3.2 Investment in Business Processes for Economic Value Added

Four factors play a key role in evaluating the economic-value-added (EVA) contribution[4] of business processes to the overall performance of the enterprise:

1. Impact on the customer
2. Relevance to overall strategic direction
3. Viability
4. Worth

3.2.1 Impact on the Customer

Impact on the customer is a central criterion for identifying a corporation's key business processes and prioritizing them for further analysis and reengineering. The ultimate measure of customer impact is the external customer's perception of the value received from the supplier or service provider. However, high customer impact does not mean that the outputs of a given business process are directly received by its external customer. This is known as indirect impact, and almost every company has a number of business processes that fall into this category. Examples include research, corporate governance, decision-making, information services, procurement, and the like.

Determining impact on the customer, therefore, may not always be easy or straightforward. Organizations cannot ask their customers directly which business processes have what impact on them and how they would rank them by this criterion. This is because many customers, even if they are somewhat knowledgeable about process performance terminology and measurement, may not have any factual information about the processes and how their suppliers use them in conducting their business. The external customer usually knows whether or not a process works, but that is insufficient for making comparative judgments.

These customers, however, are reasonably reliable in comparing the various process outcomes which they receive from their suppliers, such as product quality, price, on-time delivery, promptness of repair, accessibility of maintenance service, and correct invoicing. From their customers' comparison of process outcomes, companies can determine the relative value and, therefore, the relative impact of process outputs on their customers. These outputs can then be correlated with the business processes that have produced them to establish an importance ranking of those processes that should first be considered for reengineering.

Impact on the customer is directly related to the process performance measurement of effectiveness, which means producing a quality work product. In other words, an effective business process is one that converts input into output that meets customer requirements. Whether a customer requirement has or has not been met depends on a value judgment of the output by those who are receiving and using it — the customer, who is paying for it.

Many times, reengineering a dysfunctional business process is required to restore or improve the quality of its output, and in this sense broken or poorly performing processes should be prime targets for reengineering from a customer-impact viewpoint. Missing delivery deadlines often spurs sales and shipping to demand that manufacturing improve their processes to meet commitments.

Table 3.1. The Use of Critical Success Factors To Determine Process Relevance

Key Business Processes	Critical Success Factors (CSF) Product Quality	Supplier Quality	Empowered Skilled Employees	Customer Satisfaction	Lowest Process Cost	Lowest Service Cost	CSF Count
Weight	2	1.5	1.5	3	2	1	Max. 11
A	2	—	1.5	3	—	1	7.5
B	2	1.5	—	3	2	1	9.5
C	—	—	1.5	—	2	1	4.5
D	2	1.5	—	3	—	1	7.5

3.2.2 Relevance to Overall Strategic Direction

While impact on the customer is a powerful EVA assessment tool, it is only part of the selection criteria used by management to select key individual business processes for reengineering. Management may also want to ask whether a particular business process has a significant contribution to the organization's mission and overall strategic direction. This criterion is known as *relevance,* and its assessment is based on the approach known as the critical success factors (CSF) analysis. Table 3.1 illustrates the use of a set of strategic critical success factors in ranking the four business processes, A through D, for their relevance. Weighting is used to establish the relative importance of the CSFs within the given strategic period.

Relevant processes are among the enterprise's key processes. Relevant processes relate most directly to the firm's identity, purpose, and overall direction — those goals and attributes that visibly differentiate it from its competitors — and the priority activities that keep the company competitive. The process paradox is the result of an organization having invested in processes that were not critical to its strategic success. Such processes may have been improved, sometimes dramatically, but they did not turn out to be relevant to the company's long-term performance and success. At the same time, other key processes — the relevant ones — went unattended and sometimes even starved of resources which were shortsightedly invested in less relevant reengineering initiatives.

An example is investing significant amounts of money into the packaging process to improve the looks of packaging for a given product. If, however, product delivery misses the deadline and the products are discarded or returned to the supplier, the wrong process was improved. Instead of new packaging, what was really needed was a reduction of cycle time in both production and packaging so that the products could be shipped more expeditiously and within the required deadline.

An analysis of impact on the customer and relevance can be further complemented by the powerful technique of *competitive benchmarking,* which provides a known, possible target for process improvement. By using the results already

Table 3.2. Success Factors for Process Viability

Success Factor	*Viability Assessment*
Scope	The broader the scope, the more challenging it is to accomplish a reengineering initiative
Resource consumption and cost	Higher cost usually reduces the probability of success and may impact the potential EVA of the reengineered process
Technological risk	The newer and more complex the technology, the greater the risk of failure
Management and stakeholder commitment, support, and participation	More than half of the reengineering efforts that have failed did so because of the lack of sustained management commitment
Qualifications and competence of business process owner and the process management team	The selection of process owner and team is most important, considering the criticality of their long-term commitment to stay with the effort well into implementation

obtained by others, management can identify relevant processes and set realistic reengineering goals that everyone in the enterprise knows are attainable.

Benchmarking provides answers to such basic process-related questions as:

- Was this process found relevant in competing firms or organizations?
- Does process performance compare favorably with the best in class standard?
- Does the process represent best industry practices?
- Is the process technologically backward? Up-to-date? Advanced?

3.2.3 The Viability of the Process

The third EVA criterion is *viability*, which comprises a set of success factors that collectively determine the probability that a particular reengineering effort will be successful. The key factors that make up process viability are illustrated in Table 3.2. Until now, viability assessment has been the orphan among process reengineering methodologies. Yet, given the high proportion of reengineering projects that do not achieve their stated objectives, it is advisable for management to pay attention to these factors, because any one of them can lead to the complete failure of the reengineering initiative.

3.2.4 The Worth of the Process

The final selection criterion — and probably the most important — is the *worth* of the process. There are those business processes that return more money to the firm than they cost in terms of capital. These processes become corporate assets. Other processes cost more than what they return and, therefore, are corporate liabilities. Put simply, the worth of a process is determined by its economic value added to the corporation. An individual business process may be radically improved, resulting in impressive and measurable benefits at the process level, but these benefits may not necessarily add value to the corporation overall. Benefits such as process performance improvement, cost reduction, and the like are managerial terms and are not necessarily indicative of value added. Value is an economic concept. If the reengineering initiative costs more than its measurable benefits, then it is in essence draining the corporation's assets, not enhancing them. In assessing process costs, one must consider both direct and indirect (consequential) costs.

It is always risky to invest in processes for which the true cost is not well established, because management lacks a valid cost base against which to compare the expected benefits of reengineering. This can become another case of the process paradox: suboptimization at the process level, while ignoring the ultimate effect at the enterprise level. This lack of worth may also characterize the automation initiative where reengineering first streamlines and then accelerates, by using new technology, an administrative process that was a corporate liability. Without appropriate value-based safeguards, this reengineered process is still a liability after the reengineering and automation are completed — with the exception that it is more expensive now.

3.3 Process Cost

Historically, process management has always suffered from the lack of an obvious and reliable method of measurement that consistently indicates the level of *resource consumption* by the business process at any given time — an indicator which always interests executive management and is easily understood. The bottom line is that most businesses have no clue how much their processes really cost.

In the past, process management was not considered a part of business management; rather, it was viewed as a technical function. The reason is that, until now, processes have not been looked upon in terms of cost like everything else in the business. As part of process centering, business processes inevitably become the prime vehicles for running an adaptive enterprise. To this end, enterprises are being realigned around their high-value business processes. This systematically brings together assembled costs and resulting value estimates, which makes management listen and depend on their availability. Thus, we can identify two important results of accurately establishing process costs:

1. Gain management's attention and confidence that it can depend on them as a reliable business indicator.
2. Allow reliable assessment of process worth and its contribution to the overall performance of the business.

As we have seen, the economic value contribution of individual business processes is a major concern for management, as this information is critical to determine what key business processes the enterprise should be centered on. Before it can generate value, however, a business process must be shown to operate relatively free of waste — that is, at its optimal level of resource consumption (i.e., the process must be efficient). Consequently, one key objective of management is to minimize the total amount of resources spent on the management and operation of processes.

Operating effective and efficient processes costs money. In such a situation, affordable process costs include:

- The cost to produce defect-free products and services
- The cost to acquire and operate the information technology which enables business processes to be highly adaptable and self-organizing
- The cost to remove defects and improve products to the point where they meet customer requirements

But, it will cost much more to operate an ineffective or inefficient process. Such additional costs are unaffordable and include:

- The cost of using defective products
- The cost of having to depend on repeated service calls because of continuing poor product performance
- The cost of using erroneous reports or other incorrect information

These and other unaffordable cost elements collectively make up the waste in the process. Products that fail to meet customer requirements can cause delays, loss of productivity or assets, accidents, etc. In assessing process costs, all such factors must be considered.

While the trend of declining total process cost is, in itself, a good efficiency indicator, management is also interested in knowing if the process is operating at its optimal point of resource consumption. It is axiomatic that a process can be at its lowest level of resource consumption only if the level of waste in the process is at zero.

One of the best quantifiable measurements of waste is known as the *cost of nonconformance* (also known as the cost of poor quality, abbreviated COPQ),[5] which, in fact, is a component of the total *cost of quality* (COQ).[6] The use of the cost of quality as a process performance indicator is a classic example of adding the tools of continuous improvement to the management of business processes. The total cost of quality is traditionally divided into two components:

1. The cost of conformance, comprising the cost of prevention and the cost of predictive appraisal
2. The cost of nonconformance, which, in turn, comprises the cost of failure and the cost of detective appraisal

Typical examples for the cost elements that make up the total cost of quality are shown in Table 3.3.

Table 3.3. Typical Examples for Cost of Quality Components

Components	*Examples*	*Cost Management Categories*
Prevention	Quality education Process design Defect cause removal Process change Quality audit Maintenance (preventive)	Cost of conformance
Appraisal	Test Measurements Evaluations and assessments Problem analysis Inspection Maintenance (test equipment) Detection	Cost of nonconformance
Failure (internal)	Reinspection and retest Scrap Rework Repairs Service (unplanned) Defect removal Lost production	
Failure (external)	Returned products Legal exposure and costs Liability and damage claims Poor availability Malfunction Replacement Poor safety Complaint administration and warranty	

3.3.1 *The Cost of Conformance*

The cost of conformance is what is necessary to spend to make things come out right the first time — that is, to meet the conditions of satisfaction. It comprises the following two cost elements:

1. *Cost of prevention:* Expenditures required to prevent defects and to meet the conditions of satisfaction the first time; to minimize failure and appraisal costs. Prevention money is usually spent before the work product is actually produced.

2. *Cost of predictive appraisal:* This cost category includes all of the costs of designing and maintaining a quality system — the organization and infrastructure for evaluation and test, measurements, and so on. This appraisal money is spent before and while the work product is actually produced.

The sum of the costs of prevention and predictive appraisal is commonly known as the cost of conformance:

Cost of Conformance = Cost of Prevention + Cost of Prediction

3.3.2 The Cost of Nonconformance

The cost of nonconformance is made up of all the expenditures resulting from doing things wrong. It comprises the following two cost elements:

1. *Cost of detective appraisal*: This cost category includes all the costs incurred while conducting inspections, tests, and other planned evaluations — resulting in the actual detection of defects or variations. This appraisal money is spent during and after the work product is actually produced but usually before it is released to the customer.
2. *Cost of failure:* Money spent on activities that would not be necessary if the needs for satisfaction had been met the first time. The cost of failure further includes:

   - Internal failure costs, caused by defects that exist in a product prior to release to the customer
   - External failure costs, caused by defects detected by the customer after the product or service has been received

   Failure-related money is spent after the work product is actually produced or the service delivered; part of it may have to be spent after the product or service is received by the customer.

The sum of the costs of detective appraisal and failure is commonly known as the cost of nonconformance (also known as the cost of poor quality):

Cost of Nonconformance = Cost of Detection + Cost of Failure

3.3.3 The Cost of Quality as Management Tool

The aim of management is to minimize the total cost of quality. This can be achieved by either of the following:

1. Simultaneously reducing all four cost elements
2. Prudently increasing the cost of prevention to dramatically decrease the cost of nonconformance

Hence, the cost of prevention may increase as long as this is more than offset by the savings resulting from reduced penalties paid for nonconformance (costs of detective appraisal and failure). Any reduction in the total cost of quality usually translates directly into bottom-line profits for the business. In addition, reduction of the cost of nonconformance is the essential prerequisite for a business to operate efficiently or at its optimum level of resource consumption. In other words, *the cost of nonconformance is a lead indicator of process efficiency.*

The cost of quality is also a management tool for trade-off against other process parameters (operating costs, schedules, revenues, and so on). Its use is based on the principle that it is always less expensive to do the job right the first time than to do it over. As such, the cost of quality is one of the universal measurements across the business process, the common financial denominator which brings consistency of measurement and linkage to productivity and profits.[7]

AT&T, IBM, and many other technology manufacturers learned that in order not to miss a market window it was better to engineer an acceptable product now, working out minor bugs later, rather than to try to engineer a perfect product and deliver it too late. Such decisions were primarily based on a prudent trade-off of anticipated failure costs against the potential revenue loss of a missed market opportunity.

The largest component in the cost of quality is usually the cost of failure. To cut the cost of quality effectively, failure costs must be reduced to a minimum — if possible, totally eliminated. One way of doing this, of course, is to invest more in prevention. This financial mechanism makes the cost of quality an investment tool. What has, then, emerged here is a concept of defining and tracking conformance-related costs and then using them for two different, but interrelated, purposes:

1. Establish a tool for identifying opportunities for management trade-offs and the reduction of overall process costs.
2. Provide management with an investment tool for planning prudent increases in the cost of prevention, ultimately to achieve reductions in the total cost of quality, identified earlier.

Note that the term "cost of quality" is associated solely with customer requirements — or, more directly, with the cost of making, finding, and repairing defects — or preventing them from happening.

In a business process, however, there are many other costs which are not a part of the cost of quality and which contribute in a direct way to producing work products and to achieving customer satisfaction. These costs must also be reliably determined, because they will become part of the value calculation for process selection and alignment.

3.3.4 Productivity and Value

Reduction of waste allows the transfer of resources (people, equipment, materials, and energy) from failure-related, low-value activities to activities that produce satisfactory products and services. In effect, the reduction of waste increases the capacity of the production line and, indirectly, its productivity and added value. The link between process improvement (waste reduction) and productivity gains has significance for improved long-term process productivity. This brings us back to the original proposition of this chapter; namely, the purpose of the business process is more than producing products and services that meet requirements. The ultimate purpose is to create value for the customer and for the supplier, as well.

A work product, service, or experience need not be useful or utilitarian — artists can be productive even though their products are not used in the physical sense, but they are still highly valued. But, for work to be productive and value adding, it must, through the product or service produced, satisfy a need and expectation which makes it valuable to the recipient — the customer. Thus, the measure of productivity of a work process is the value added to the products and services produced, divided by the value of the resources consumed.

Value is usually determined by the marketplace, or even by individual customers, through the laws of supply and demand. Value measures the very requirements against which a product or service is provided; it is inherent in the agreed-upon conditions of satisfaction. Two providers can produce identical products, but one provider is more productive than the other if its products are in demand (they are perceived to have more value) and the other's are not. In other words, a company producing what the market requires is more productive than one producing unnecessary items, however well they conform to some specification or resource utilization formula. This is the essential difference between requirements and specifications — requirements are always based on a stated or recognized customer need and expectation.

Efforts to improve productivity are often hampered by these widely held misconceptions:

- The solution to the productivity problem is increased capital spending.
- Productivity gains are difficult to achieve in service or white-collar industries.
- The problem with productivity revolves around the workforce.

In fact, to have any major impact on productivity, management action is required. The vehicle for achieving any measurable change in productivity is improving the business process. To define productivity measures and then improve them, it is necessary to assign a value to each product or service produced. These are then arranged into an optimal pattern through appropriate alignment of the business processes responsible for producing them. With value determined for a given point in time, productivity gains can arise only through adjusting either the business process or its inputs. Generally, adjustments to the business process yield the greatest long-term gains.

The necessary, but not sufficient, condition for productivity improvement and reduced costs is the reengineering of individual business processes. Reengineering also generates EVA, if the reengineered process returns more money to the enterprise than it truly costs. For providing maximum value and then achieving continuous value growth by the enterprise, the sufficient condition is the alignment of the enterprise around its key value-adding processes. The two together add up to process centering.

In the emerging sense-and-respond[8] world of business, sensing a need and then responding to that need by conforming to the resulting new requirements (that is, meeting conditions of satisfaction) is one way of creating value. This is also known as *passive adaptation* to external change.

Another way is to recognize opportunities for improvement (generated by new technologies, changes in the financial or monetary environment, regulatory or legislative changes, etc.) and initiate a change in response to that opportunity. The change will lead to business action, such as the discovery of new uses and new markets, formulation of new product and service requirements, and so on. This is known as *active adaptation*; it is self-initiated and only indirectly induced by the environment.[9]

Agility is a unique manifestation of active adaptation. It is the ability to grow and prosper in an environment of accelerating and unpredictable change — by recognizing and profiting from shifting customer needs and expectations, and from opportunities for offering customized, value-based products and services.

3.4 Summary

After long periods of careful process redesign, many companies have achieved impressive improvements in individual business processes, only to watch overall business performance decline. The reason for this outcome is the process paradox — companies have invested in the reengineering of business processes that do not add economic value to the overall performance of the enterprise.

More recently, management discovered that business performance can further be improved by the reengineering of its core processes: the aggregations of customer value-adding work. Four factors play a key role in evaluating the economic-value-added (EVA) contribution of business processes to the overall performance of the enterprise: impact on the customer, relevance to overall strategic direction, viability, and worth.

The economic value contribution of individual business processes is a major concern for management, as this information is critical to determine what key business processes the enterprise should be centered on. Before it can generate value, however, a business process must be shown to operate relatively free of waste (the process is efficient). Consequently, one key objective of management is to minimize the total amount of resources consumed by the management and operation of processes (i.e., minimize process cost). The cost of quality is one of management's tools in containing and minimizing process cost. It is also a classic example of adding the tools of continuous improvement to the management of business processes. The ultimate measure of productivity of a work process is the value added to the products and services produced, divided by the value of the resources consumed.

4 The Intellectual Value in Process Centering

4.1 The Emergence of Intellectual Assets

The existence of the traditional firm is equated to the ownership of a collection of finite *things*, usually representing tangible items, such as land, equipment, plant, and other material assets, financed by shareholders. The traditional view of the firm and the things it owns reflects the *physicality* of assets.

In today's information age, the collection of things comprises more the intellectual assets the firm owns: brands, knowledge and skills of its employees, and the know-how (information and knowledge) accumulated over time. They reflect the *intellectuality* of assets, financed by client investors as opposed to shareholder investors. In other words, investment is not made in terms of cash or capital or other tangible assets; it is made in terms of the transfer and accumulation of information and knowledge

At General Electric, about half of productivity gains attributed to the company now come from investing intellectual capital (that is, research and development) rather than real capital. In the pharmaceutical industry, this ratio is even higher today. During the last decade, 53% of U.S. companies' capital budgets has been spent on equipment handling information rather than raw materials, energy, or other material objects.[1] In many companies, physical assets that used to be the repositories of raw materials, semi-finished goods, parts inventory, and the like are now used for generating intellectual assets, such as new skills and knowledge (e.g., training centers and development laboratories, among others). J.P. Morgan's former vault, once home to gold bouillon, cash, and securities, is now their training center at 60 Wall Street in New York City.

As we have already discussed in the previous chapter, traditional economics have difficulty dealing with the value of knowledge, information, innovation, and management ideas. Nevertheless, we know that in the case of established companies the market knows how to value innovation (e.g., research and development) and the intellectual assets they produce. It has also been recognized that the investment of intellectual capital yields superior returns as compared to the investment of financial or physical assets. While the average after-tax return on physical assets in the U.S. has been a little over 4% in the 1990s, the return on intellectual assets has been valued at over 16%. Information technology, which is designed to handle and transform intellectual assets, is the fastest growing sector of the gross domestic product.

The challenge in managing business processes and, in particular, the process-centered enterprise is to identify intellectual assets as part of the process capabilities. Just like physical process assets (e.g., materials, equipment, and energy), intellectual assets are not inputs to the process; they are transformative mechanisms that can transform physical inputs or information and knowledge into desired process outcomes.

When designing a bridge, a good structural engineer mobilizes several sources of knowledge:

- Formal education in structural design, methods, and tools
- Standards and building codes
- Technical requirements and specifications
- Capital and expense budgets
- Legal and regulatory codes
- Insights and experiences of fellow engineers, academics, and other professionals

These various sources of knowledge represent intellectual assets. The engineer applies them in the design process to transform physical inputs, such as structural steel, concrete, masonry, timber, and high-tension bolts, into the design of a load-bearing structure that will serve the ultimate client's needs. Without these intellectual assets, a very different structure would be produced or, most probably, none at all. Of course, the construction process for building the bridge also utilizes both physical assets (e.g., the construction materials, construction machinery, scaffolding, transportation equipment, and so forth) and intellectual assets (the bridge design, construction methods, and expertise) to transform the same inputs into the actual bridge — as the final outcome.

Critical intellectual assets of a business process are, of course, the *commitments* which exist at any given time and which govern the execution of all process activities. The decisions leading to these commitments, as well as the information and knowledge used in the decision-making process, are also part of the intellectual capital. As we will see later, identification and definition of intellectual assets are just as important in process design as the creation of process maps and the definition of work activities, along with the tangible resources necessary to perform them.

The starting point for managing intellectual assets in a business process is an understanding of *core information assets.* Core information assets represent a competitive advantage for the enterprise. They have been developed over a long period of time and cannot be easily acquired or imitated. They are distinct from tangible, energy- and matter-based assets, which are not necessarily superior to those of competitors and therefore do not offer a significant competitive advantage. Examples include inventions, patents, and trademarks; operating procedures and work practices; customer databases — just to name a few.

The idea that certain intellectual assets bring competitive advantages to a company led academic researchers and business executives to the recognition that a strategy building on distinctive intellectual assets can lead to superior competitive performance in the marketplace. Building certain intellectual assets and then developing process capabilities for exploiting them seems to be the strategy of the moment.

4.2 Intellectual Value Added

The old adage, "What gets measured, gets done," also holds true for the value of intellectual capital. There are many reasonably consistent and meaningful ways to measure the components of intellectual capital and to identify the business processes that can enhance it. For example, dollar values can be assigned to such things as brand equity, customer loyalty, or the customer base in general. Other metrics do not attempt to ascribe a dollar figure to the value of intangibles — such as human capital, intellectual assets, and newly created knowledge. Rather, they measure the performance of processes and their results that depend on those intangibles.

While knowledge itself is difficult to quantify, both the process activities using knowledge and the process outputs resulting from its use can be measured. For example, a pharmaceutical firm can quantify what knowledge is being used (e.g., number of patents, publications and articles, the number and quality of validated tests, and the number of results generated by previous experiments) and what contribution the knowledge has made to the drug-development process itself (e.g., the number of drugs brought to market, reduced cost and cycle time, increased revenue, among others).

Adaptability of business processes and the responsiveness of the enterprise depend on speed, innovation, and flexibility. Innovation is the creative and timely application of information and knowledge to a heretofore unresolved problem; flexibility is the ability to rapidly change and realign commitments or create new ones — in the face of change.

Many companies, both in the manufacturing and the service sector, regularly track the percentage of revenue generated by new products and services (that is, those launched within a pre-determined timeframe, say, the last two years). This is certainly an insightful measure of innovation and of the ability to translate the results of research and development into commercially successful products and services — all part of the organization's intellectual capital. Hewlett-Packard is one among several technology companies that consistently track the percentage of sales generated by new products. IBM has just recently mounted an effort to practically mandate the amount of time its research scientists must spend with customers — to stimulate innovation and the anticipation of future needs.

Three approaches can be used to measure intellectual assets and the processes that either use them or contribute to them:

- Customer capital, as represented by customer loyalty
- Intellectual capital
- Net added value of information processed (databases or communications)

4.2.1 Customer Capital

Customers are the most important external asset an enterprise can have. When a company is the preferred supplier to a financially strong customer, that customer's loyalty can be an invaluable economic asset. The value of this asset can be at least approximated by finding the answers to some relatively simple questions:

- What is the financial strength and growth potential of our customers?
- What is our share of our customers' total supplier expenditure?
- What is the present net value of our customer base?
- How much is a loyal customer worth?
- How much is it worth to keep a particular customer?
- How much is a new customer worth — after subtracting the investment to attract them as a customer?

Study after study has shown that customer retention is the best possible growth strategy for an enterprise. Losing a customer is very expensive. One approach to evaluating how much a customer is worth is as follows:

1. *Determine the time period over which the evaluation will be done.* This varies greatly by the type of business, the length of the business cycle, and the like. A commercial lender would track customers for decades because it takes more than a few years to realize good returns on capital loans. A credit card company may track a customer's worth for a few years because spending habits and credit worthiness require an assessment over two or three years. A commercial laundry, however, would only need to track a customer for a year or less, because its volumes are seasonal and can be determined over a shorter period.
2. *Calculate the annual profit typically generated by representative categories of customers.* These include newcomers, loyal customers with long histories, and customers currently not satisfied with the products or services they receive. In calculating profit by customer, it should be remembered that costs could vary significantly from category to category and even for individual customers. On the revenue side, long-term loyal customers tend to make

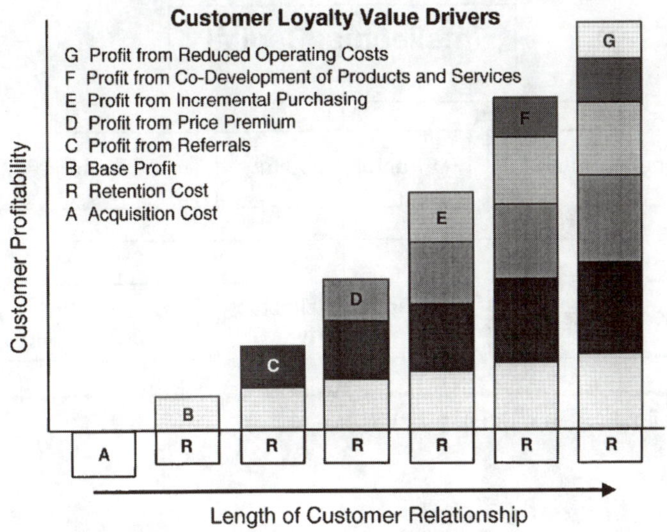

Figure 4.1. Customer Profitability

prices, buying multiple products and services, and generating additional business through referrals (Figure 4.1).

3. *Determine customer retention,* which is the quantification of the net remaining customer base each year, after erosion due to customer disloyalty (a customer taking business elsewhere).
4. *Based on the annual profit per customer and the customer retention measures, calculate the net present value of each customer.* Because customer capital is considered an asset, a discount rate can be selected that represents the desired annual return on assets. This discount rate is then applied to each year's profit, adjusted for the likelihood (derived from the customer retention measures) that the customer would become disloyal and leave. The sum of these adjusted annual amounts over the evaluation period is how much the customer is worth — the net present value of all the profits that can be expected. In other words, it is the amount that a competitor would want to pay to lure away that particular customer.

Armed with this information, management can make better investment decisions as to how much to spend on attracting new customers and which business processes will contribute the most to it, which customers to pursue and which not to, and how to exploit the economic leverage of long-term customer loyalty. One key result of customer loyalty is repeat business; this is also the ultimate manifestation of customer capital. Its ongoing enhancement usually merits bigger investments than many companies tend to make.

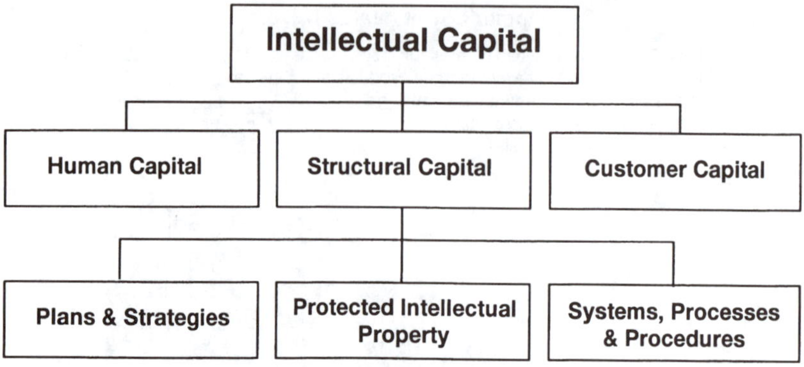

Figure 4.2. Intellectual Capital

4.2.2 Intellectual Capital

It is customary to identify three components of intellectual capital (Figure 4.2):[2]

1. *Human capital*, which is essentially the same as the intellectual resources described in Section 4.1 above. Among its quantitative measures are the identification of skills and competencies most valued by customers or most sought after within the enterprise.
2. *Structural capital*, which includes (a) intellectual property protected by patents or copyright; (b) systems, processes, and procedures of the organization; and (c) plans and strategies.
3. *Customer capital*, in terms of customer loyalty, which was discussed earlier in this section. Another component of customer capital is customer satisfaction, a prerequisite of loyalty. Customer satisfaction can be measured in absolute terms, as a trend over time, or as an economic asset.

4.2.3 Net Added Value of Information Processed

In a process-centered environment, the processing of information essentially boils down to the execution of these four steps:

1. Sensing and interpreting requirements and change
2. Deciding on the appropriate response
3. Making the necessary commitments for implementing the response
4. Describing and tracking the procedural activities committed and executed to produce the desired result

The basic forms of processed information include:

- Formal communications, such as requirement statements, requests for proposals, offers, negotiations, and so on, which are necessary for reaching decisions and making commitments; this form of information can also be characterized as *commissive*, because it eventually leads to making and managing commitments

- Decisions,[3] such as agreements and commitments regarding conditions of satisfaction, deliverables, customer acceptance, and its various forms[4]
- Process-related descriptive information, such as the process plan, process maps and models, systems requirements, organizational design, performance measurements
- Product-related prescriptive information, such as designs, patents, models, drawings, maintenance, and service instructions, which is related to the process output received by the customer
- Supportive information, such as education and training, instructions for use, and safety warnings, which is related to the product or service

The net added value of processed information lies in the potential of transforming that information into marketable products and services. The value of a good proposal will be reflected by the contract amount won. The design of a car is ultimately valued by its contribution to market share growth, and so on.

4.3 The Role of Knowledge Management in Process Design

Knowledge is defined as interpreted information that has been validated to be true, thus it can enter the body of general knowledge accumulated over time — be it scientific, economic, social, or political. In any enterprise, business processes both use and generate information and knowledge. The sensing, capture, distribution, storage, and retrieval of information and its conversion into knowledge are the tasks of knowledge management. Knowledge management also includes the development of knowledge-related strategies and the establishment of an appropriate management infrastructure.

Information and knowledge can be both enabling and integrating assets. When designing critical business processes for a given organization, the traditional reengineering movement paid little attention to this fact. Accumulative knowledge about customers, suppliers, and, most critically, the test performance of the process itself has not been addressed in a process design in any systematic way. As a result, reengineered processes have occasionally developed a "knowledge gap" and ended up devoid of intellectual assets.

For example, the mechanistic redesign of a business process often results in a reduction of work steps and hand-off points — with an attendant reduction of related jobs. Jobs, however, are occupied by people who, in turn, represent a certain specialized know-how and knowledge. Their redeployment or retirement will inevitably result in a loss of intellectual assets which some firms remedy by rehiring former employees as consultants, often at a much higher price. In any case, these post-reengineering efforts to salvage and recapture knowledge are expensive and ineffective in the long run.

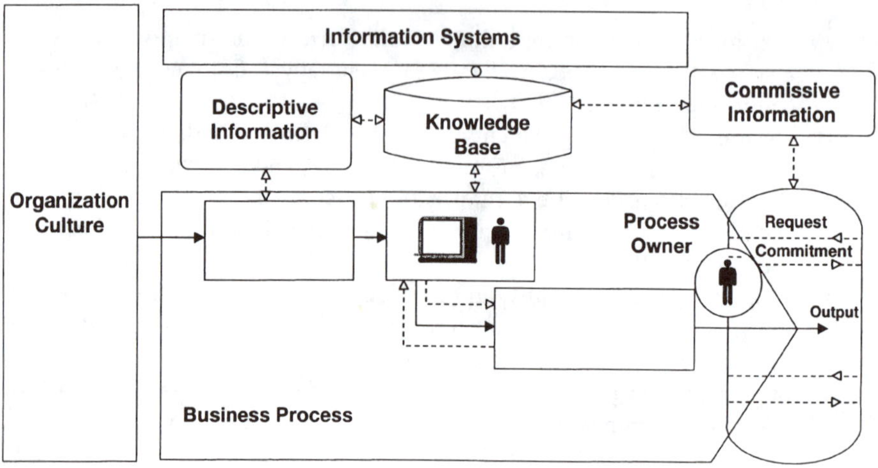

Figure 4.3. Process-Centered Information Architecture

In today's world of accelerating and unpredictable change, enterprises are beginning to understand how certain forms of information and knowledge can be leveraged not only to improve their current performance in terms of process efficiency and effectiveness, but also to enhance the adaptability of their processes in terms of ability to reassess and change commitments and to find innovative responses to change. Business process redesign no longer focuses solely on taking out non-value-adding activities of the process but also on maximizing the knowledge content and expertise of their employees to improve organizational performance.

As we will see later, intelligent process design requires that information and the knowledge are leveraged to improve process performance. This in turn requires an information architecture for organizing the capture, share, and use of information and knowledge in the day-to-day operation of the process. Figure 4.3 illustrates the process-centered information architecture associated with an operating business process. The process owner is focused on decision-making in the face of incoming customer requests and change — using and creating commissive information and adding the resulting new knowledge to the base. Knowledge workers in the process both use and generate what is primarily descriptive information — which eventually also accumulates in the organization's knowledge base.

All decision-making and the management of commitments in a business process depend primarily on current information and existing knowledge. Every time a procedural work activity is executed, existing knowledge is used. At the same time, with every decision made and every commitment met, existing knowledge is enhanced or new knowledge is generated in the process. In most of today's organizations, however, enhanced knowledge is seldom captured for future benefit and, as a result, rarely applied during the next instance of executing the same process.

Certain complex processes, such as new product introduction, depend on the knowledge and experience of literally hundreds of professionals and managers at very different points during the process cycle. Many of these individuals are the key decision-makers in the process, who also make commitments and are held accountable for successful delivery against them. In their decision-making, they use knowledge that often is critical and time sensitive and must be kept current and up-to-date at all times.

Identifying these accountable individuals is extremely helpful in bringing relevant knowledge to the decision-making and commitment management process, which will then assuredly lead to informed decision-making in the face of change. We will see later that the alignment of these key individuals in the process occurs along the customer-supplier value chain. In fact, this set of accountable individuals manages the network of commitments. In doing so, they conduct an ongoing customer-supplier dialogue, continuously using all the available information and knowledge in arriving at decisions and agreements and making commitments.

Ultimately, process design must ensure that information and knowledge in the system are used to get better work done in less time and at lower cost. Knowledge management in itself is not value; value comes from creating new, tangible, and intellectual assets with it.

4.4 Summary

In today's information age, the collection of things more typically involves the intellectual assets the firm owns, such as brands, knowledge and skills of its employees, and information and knowledge accumulated over time. These intellectual assets are financed by client investors as opposed to shareholder investors. In several industries, over half of the productivity gains now come from investing intellectual capital.

Just as for physical process assets, intellectual assets are not inputs to the process; rather, they are transformative mechanisms that can transform physical inputs or information and knowledge into desired process outcomes. In other words, both information and knowledge are enabling assets. In a process-centered enterprise, individual business processes can contribute added intellectual value by enhancing the intellectual capital of the enterprise. Key to these enhancements is the ability to assign some financial measure of worth to customer capital and to measure, in some quantitative way, the various other asset components of intellectual capital.

Enterprises are beginning to understand how certain forms of information and knowledge can be leveraged not only to improve their current performance in terms of process efficiency and effectiveness, but also to enhance the adaptability of their processes — in terms of ability to reassess and change commitments and to find innovative responses to change. It is the task of knowledge management to establish the strategies and infrastructure necessary to build and maintain an enterprise's intellectual capital base, and to design them into individual business processes — and across the enterprise.

Part II: Process-Centering Fundamentals

5 Understanding Processes

Change is a fact of life. External change requires internal change implemented in a number of ways: one-time downsizing, continuous improvement, process reengineering, and so on. In the face of change, the view that "If it ain't broke, don't fix it" (that is, doing nothing) could spell disaster. Though many organizations seem to be doing well, most begin to realize that failure to change significantly in the next few years will result in being noncompetitive in the long term. Because those doing well have no immediate motivation to change, they are the most vulnerable.

The shock of watching giant IBM, a traditionally organized firm, falter in the early 1990s has left many boards of large corporations scrambling for better ways to deal with an increasingly competitive and unpredictable environment. The central question for us, then, is this: What organizational form and what management approach will be more successful in ensuring the survival and competitiveness of the enterprise?

Internal change can be either passive or active. In response to external change — expected or not — passive internal change can occur spontaneously in any part of the organization and at any level. The problem with passive change is that, when and if it does take place, it is largely uncoordinated and often inefficient. Active change, on the other hand, is by definition always management induced; its vehicles of implementation are the organization's business processes. To understand how to harness these processes for the task, a review of a few basics is necessary.

5.1 Process Fundamentals

In the discussion that follows, we will first use classic definitions and terminology to help the reader understand process basics and to establish the foundations of the world of business processes.

5.1.1 Classic Definitions

While many texts and authors use the terms *system* and *process* interchangeably, there are some fundamental differences which need defining:

- *System:* the standing logical organization of living organisms, methods, mechanisms, and automatic devices which are interrelated and defined as a whole to accomplish a specified result (objective).
- *Process:* the logical set of living organisms (usually people), materials, equipment, energy, and information organized into work activities to transform given inputs into a specified output (result).

The classic definition of a system is made up of three basic tenets:

1. It is a logically organized set of entities that interrelate with each other.
2. As a whole, it is standing and ready to perform work.
3. It cannot be divided into independent parts.

A process is a system put into operation to produce a specified result. This definition of a process comprises two essential tenets: (1) inputs are known or predictable, and (2) work is being performed by resources over time to transform input into output (the result).

As we will see later, inputs, such as customer requests, are not always predictable; in fact, in today's fast-changing world, they are becoming highly unpredictable. How a process deals with this situation is the subject of the chapters in Part II of this book and is our basic message. For its performance to be manageable, a process must be measurable, which means it must:

- Have defined inputs
- Be stable during the period of observation
- Have predictable outputs

Once a process is deemed manageable, its performance can be measured in terms of effectiveness and efficiency. An effective process produces output that meets customer requirements (or, which is the same, agreed-upon conditions of satisfaction); an efficient process does this with minimum resource consumption.

The resulting work product must have a value (measurable in some appropriate form such as price or other value indicator) greater than the value of the original inputs to the process. In other words, the process must *add* value to the original inputs as it converts them to outputs, which is the classic input-process-output (IPO) definition of a business process.[1] A good example of this is the assembly line in an automobile factory. If the cost of all the parts and materials, labor, and capital invested in making a car exceeds what it is sold for, then the process adds negative value. It should be improved or shut down.

5.1.2 Process Control

In the classic world of processes, the ideal state to be in is one that is stable. Process control, one of the most important tools of process management, has a single purpose: to maintain the process in a stable state or return it to stability once the original state is disturbed by variation or change. In a stable world, all processes are definable and repeatable, plus their performance is predictable. The performance of a stable process includes not only the final result, but also any intermediate results that indicate the operating state of the process, such as elapsed time, cost, yield, flow rates, and so forth, indicating the amount and rate of resource consumption in the process.

The world of business, however, is dynamic and full of change — some predictable but most of it increasingly unpredictable. For decades, process managers have been trying to create systems to maintain the *status quo* rather than accept the instability of the world and devise new processes to deal with that instability. In the past, only in emergencies or times of crisis did management try to adapt to the change forced upon them. Interestingly enough, entire manufacturing industries were transformed in record time during World War II to meet the new demand for war planes, warships, weapons, and other needs of the military. These demands were very different from the normal course of business; yet, given no choice, they were met.

Today's onslaught of accelerating and unpredictable change is similar to war-like conditions; these changes leave use no choice but to adapt or face the consequences. This new business environment demands a change in process design — processes must be more adaptable. In adaptable processes, process control takes on an added dimension: the assurance of progression from one stable state to the next.

As the real world is not stable, an enterprise's major strategy for future success must be to develop processes that can adapt to constant change and instability. In these processes, people would not only accept the inevitability of frequent change but would welcome it. They see changes bring energy, variety, challenge, opportunity, greater empowerment, and the knowledge that somebody appreciates their knowledge and skills to deal with change and the inter-connectedness of their work.

5.1.3 Process Capability

Understanding and applying process capability is one of the basic tenets of this book. Process capability is defined as *the measure of the inherent uniformity of process performance.* Practically speaking, a process is termed capable if it is able to consistently achieve a predetermined pattern of results (e.g., meet requirements such as all washing machines having no more than five service calls over a 20-year use life or a carton of milk having a 5-day shelf life).

With unpredictable change, the dynamics of the situation change. Requirements placed upon a process will change on an ongoing basis, so its capability has to be constantly reconfigured and adapted to the new requirements. While process control is an instrument of stability, process capability is the instrument of change and adaptability.

A process is termed adaptable if it is able to maintain its capability or achieve improved levels of capability in the face of change while maintaining its measured effectiveness and efficiency over time. The bacteria that have been developed to help contain oil spills do just this. The more oil they eat, the more they multiply, so the more oil they can eat. Rather than losing their effectiveness in the early stages of an expanding oil spill, they become dynamic and maintain their effectiveness and efficiency. As a self-enhancing ecological process, they offer a revolutionary new tool for dealing with contamination from oil spills.

Maintaining stability is analogous to the situation where someone is trying to plug an increasingly leaky dam. Why not let the water through into a container pond and make use of it for productive purposes such as irrigation? Other examples:

- During the floods along the Mississippi several years ago, land was purposely flooded to save other areas downstream. Initially, many misunderstood what was being done, but the process saved tens of thousands of acres of arable land. Flooding some land to save other land saved millions of dollars in damage in the long run.
- Firefighters allowed thousands of acres to burn in Yellowstone National Park several years ago, and the public was shocked; however, the new vegetation has restored the health and beauty of several major areas that had decayed over the years.

This is the fundamental philosophical question behind response to change. Rather than resist it and try to maintain a state of permanence and stability, why not actively sense, interpret, and then capitalize on change for one's own benefit? Thinking in these terms has sometimes been uncomfortable because it means short-term sacrifice may be inevitable, but for the mature business, the issue is about long-term survival as it continuously creates and identifies new opportunities.

Change may be unpredictable, but it is neither random nor incoherent because it is always the result of outcomes generated by other processes. So we can conclude that stability is only a temporary state of what is going on; what is the norm and what endures is the adaptable process — open, dynamic, and responsive. The dynamic world of accelerating change calls for adaptable, self-organizing structures, increasingly open systems which ultimately respond to change by creating changes in and outside themselves, as well.

5.2 Core Concepts of Process Thinking

In a functionally organized company, the flow of work is fragmented into specialized entities called *functions*, such as marketing, development, production, sales, or distribution in a manufacturing company. Work is, therefore, managed within the various functions in the enterprise. Underlying functional management is functional thinking, which has its roots in Adam Smith's basic concept of division of labor described in Chapter 1. Today most managers are functional thinkers; tomorrow they must be process-centered thinkers so enterprises can adapt and stay competitive. To be competitive in the 21st century demands that adaptable business processes become the basic mechanisms of management. This is the key to process centering. As a prerequisite, process thinking must be developed both individually and organizationally.

Process centering requires fundamental and pervasive process orientation throughout the enterprise. The organization must change its fundamental outlook and behavior. Because most people are oriented to tasks, jobs, functions, organization, and products, but not processes, a process orientation may be difficult at first.

Process thinking helps define the process environment and establish a process orientation in the business. Process thinking requires a change from thinking in pieces to thinking about a dynamic, fluid system in which all of the parts are affected by any decision or change in the other parts. This is the type of thinking employees of the 21st century enterprise should be taught. They need to be taught to think and to understand all parts of the business so they can provide management with meaningful feedback and suggestions. The good news is that newly available software programs help map these processes and help people understand how one function affects the dynamics of every other part of the system. For example, one of the key values of enterprise resource planning (ERP) software systems is that they really help people understand business processes.

Think of an enterprise as a pond whose surface ripples when someone throws a rock into it. The rock represents changes in the environment — new requirements or new opportunities. Even though the rock impacts only one spot on the surface, eventually the entire pond will be affected in some way. Another way to look at this is to think about what we have learned about the world as a global system. At one time, we did not think that what happens in the Middle East, Malaysia, Indonesia, Russia, or Kosovo would impact life in the United States. However, in recent years, we have learned unequivocally that political turmoil in remote corners of the world eventually affects the economy and security of the United States. We can no longer take a local view of world politics. The same is true in the money markets of the world. What happens in one part of the world now instantaneously affects what happens on Wall Street — as witnessed by the collapse of the Asian and Russian financial markets in mid-1998. Using isolationism to safeguard domestic stability is no longer a viable option.

In business, our thinking is often tunnel-visioned; we are so function oriented that we do not understand the larger picture and the broader implications of our actions. Some naively believe that what each person is doing may exist in a vacuum, or our perception is warped by the limited information we have. This is functional thinking. The first step is understanding the fundamentals of process thinking and how it differs from functional thinking. Process thinking is based on three core concepts: subject process, process feedback, and process quality.

5.2.1 Subject Process

Any process is always part of a complex, larger set of processes. A *subject process* is merely that part of the larger set of processes on which attention is focused at any particular time. When a particular process is singled out for review, analysis, or reengineering, by definition it becomes the subject process. An example for this is presented in the next section.

A subject process is defined with respect to its environment by an appropriately (arbitrarily or methodically) selected boundary. The boundary, in turn, is made up of one or more interfaces defining unique relationships between the subject process and other processes. This is important, as it allows one to think in terms of networks of related processes (Figure 5.1), like business units or an enterprise connected to its suppliers and customers. It also helps define customer-supplier relationships that link various business processes to each other. Last, but not least, it enables managers to begin changing locally before company-wide changes are instituted.

As we examine a process — the subject process — we find it displays a number of attributes and behaviors. The sum of it is that the process receives inputs from

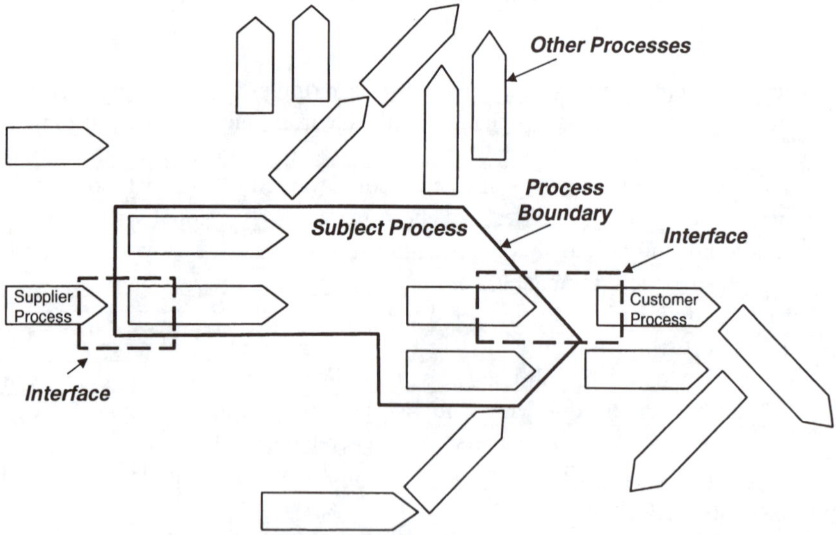

Figure 5.1. The Process Environment

supplier processes in its environment; through work activities, it transforms those inputs into outputs that are then passed to *customer processes* in the environment. There are a number of specific rules and attributes that completely describe the workings of the subject process.[2]

5.2.2 Example for Subject Processes

To illustrate process thinking, let's examine Pfizer Central Research Division's recently reengineered procurement process as a typical subject process.

1. The supplier processes providing various inputs to the procurement process are the hundreds of vendors and suppliers from whom Pfizer purchases a variety of products and services. (One key input to traditional purchasing operations is the purchase order. This, however, has been largely eliminated in Pfizer, as most employees are now authorized to initiate their own purchases online.)
2. The customer processes are the scientific research and related administrative work processes which utilize, in various forms, the items purchased.
3. Moving into the procurement process reveals many smaller processes: subprocesses, activities and tasks such as looking up items on approved vendor lists, completing purchase requests, reviewing suppliers' invoices, making payments to suppliers, and the like.
4. The boundaries of the procurement process can be established as being the receiving function for inputs (incoming items already purchased) at the input end and the stockroom near the recipients' work areas at the output end.
5. The output of the procurement process is represented by the purchased items delivered to the originators of the purchases (or to a stockroom near their activity). Any purchased item that is unacceptable to the recipients or users must eventually be returned to the supplier and considered waste. If this happens often, purchase requests to the suppliers of unsatisfactory items (inputs into their processes) will eventually cease, as Pfizer will look for alternative sources for those items.
6. Outcomes of the procurement process are effects of purchased items further downstream from the original purchaser of a given item. For instance, reagents purchased for certain toxicity tests may help reveal carcinogenic effects in a drug under development. This is a beneficial outcome (not visible from the procurement process), because it allows scientists to modify the underlying chemical formulation so that when it is brought to market, the drug will not contain a carcinogen and will be demonstrably safe in this respect.

5.2.3 Process Feedback

The second core concept of process thinking is that of *feedback*. Feedback is information about the subject process or its output which, when fed back into the process,

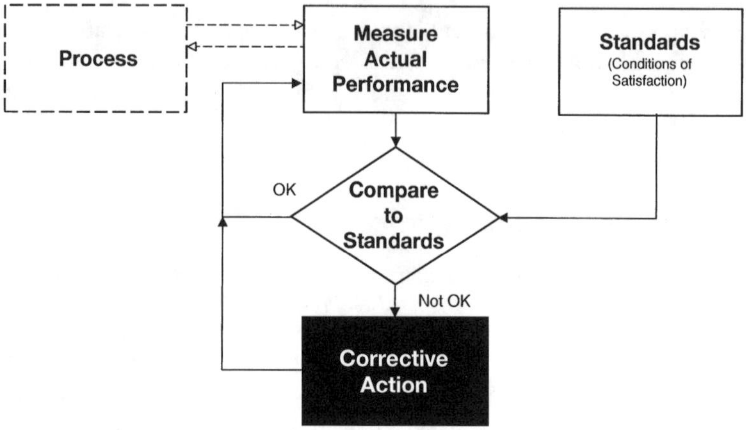

Figure 5.2. Feedback Loop

helps modify it either after its core activities have been completed or while they are still in progress — it is this case where feedback makes the process adaptable.

The basic mechanism of feedback is the feedback loop, also known as the feedback learning loop (Figure 5.2). The more effective the feedback loops are in a process, the more adaptable the process will be when there is a need to change. In many organizations right now, there is no evidence of using feedback to modify a process during execution. As a result, completing the process becomes more important than making sure it is capable of adapting to changing customer needs — a convenient attitude for the process management, but very customer unfriendly.

A closed process is one with no feedback loops; once started, process activities continue to completion. Such a process is more appropriately called a procedure. An example of a closed process is what happened to the U.S. Postal Service when they instituted Priority Mail. In the beginning, the Postal Service promised two-day delivery. What happened was that some pieces arrived in one day, but some took as many as five days in more remote locations in the U.S. As a result, some companies lost business when their bid packages did not arrive by the deadlines set. The Postal Service received a torrent of complaints when the public decided to hold them to their promises. The resulting service headaches were enormous. As they were locked into their advertising campaign, printed labels, and training, it took months to correct the problems and stop promising what they could not deliver. They had no effective process for handling the feedback, even in very serious, litigation-inducing cases.

On the other hand, an open process is one with effective feedback loops — process activities can be modified while in progress (self-corrective process). Open processes are also the base of purposeful behavior, which means they are capable of making choices and arriving at decisions.

Federal Express, for example, can tell its customers where their letter or package is in the system, as well as when it was delivered and who signed for it. Because of a sophisticated label tracking system, tracing errant items is very manageable and is an essential part of the service.

5.2.4 Process Quality

In any business process, quality output is defined as one that meets agreed-upon conditions of satisfaction. In this age of consumerism, those with the buying power will always demand quality products and services. Quality is a general expectation, no longer a value-adding feature. Quality is meaningful only when the supplier knows exactly what the customer needs; that is, when a set of agreed-to conditions of satisfaction exist. Quality means customer satisfaction, a necessity if a business wishes to survive, let alone be successful. Quality becomes part of process thinking through the laws of process quality.[3]

Whether they buy a Saturn or Mercedes, people expect the car to be free of defects and to function as advertised. On the other hand, a malfunction or defect in either car will be viewed as poor quality and cause dissatisfaction with the product. Few people will brag to their friends if their car is working as it should, but they will bitterly complain about defects and malfunctions — and the dealer who sold them a lemon.

Generally, people perceive buying a car with fabric upholstery as satisfactory, but buying a car with leather upholstery is considered buying a quality feature that further enhances satisfaction. The ritual of delivering a car to a customer in a personalized ceremony at Saturn adds a further sense of quality — even though the event costs very little relative to the overall cost of the vehicle.

5.3 The Concept of Social Processes: The Human Element

Early reengineering books were heavily criticized for ignoring the human element in the various process reengineering approaches they advocated. For an organization to be truly process centered and adaptive, the key element in its success is the people who work in its processes. In recent years, businesses have given lip service to the idea that employees are the most valued resources of an enterprise, but those wishing to transform their organization from functional hierarchies to process-centered enterprises will have to understand why the human factor is so critical to future success. Valuing human capital has come of age.

Entrepreneurial businesses frequently succeed in daunting circumstances. The employees are overworked, underpaid, and short on resources to do their jobs, but often receive stock options in lieu of cash compensation. Oftentimes, they perform brilliantly and tirelessly. Recently a California-based software company named Kington was sold; the owners, in order to say thank you to their employees for their contribution to the value of the company, shared the financial windfall of over $100 million. The average amount paid to each employee was $70,000.

5.3.1 Open Systems

An open system is made capable of self-renewal by its human component. It is also capable of dealing with change in the form of incoming information and of responding to change by creating capabilities within itself that are in turn able to attain higher levels of performance. Open systems, in which humans have a decisive role, are known as social systems; when they become operational, they turn into social processes.[4]

A lot of customer service units within major businesses now include processes that allow their employees to make decisions on the spot to take care of unexpected customer needs (within reason of course). A Federal Express story that is legend in the customer service arena is one in which an employee drives all night in his own car to deliver a critically needed package to a long-time Federal Express customer. The employee was rewarded and lauded for "breaking the rules" to fulfill what Federal Express feels is its mission: helping its customers succeed.

5.3.2 Business Processes

As we have said earlier, the traditional input-process-output model views the process as a set of repeatable activities, transforming inputs into outputs. However, this is an incomplete view because a business process is not just a repeatable procedure; it is a highly information-intensive, open system in which empowered people continuously adapt the flow of information and work to newly sensed change.

We have to be careful, then, to distinguish between repeatable, deterministic processes, which we will call procedures, and all other forms of social processes, such as communication processes, *ad hoc* activities, negotiations, meetings, and the like. Accordingly, this is the more complete definition for a business process we will use throughout this book:

> *A business process is a logical set of customer-supplier relationships that drive work activities performed by people and other resource-consuming assets in order to produce committed results at specified points in time.*

This *holistic process model* is rooted in the original work of Flores, Scherr, and others.[5] Its basic tenets are

- Inputs are often unpredictable.
- Process work is driven by commitments made to customers.
- Commitments are made by empowered, accountable people who are a part of a series of customer-supplier relationships.

A process thus defined is dynamic and, by definition, adaptable. This definition allows us to deal with traditional industrial production processes whose output is a physical product such as car parts, computer chips, airplanes, or garments. Most importantly, however, this definition is eminently applicable when the product is information or knowledge, including decisions, negotiated agreements, regulations, policies, guidelines, and laws. In the later chapters, we will cite several examples of this information-intensive approach.

5.4 Summary

The response to external change is internal change. Active internal change is management induced; its vehicles are the organization's business processes. The enterprise will be optimally responsive to change once it is process centered, aligned around its adaptable core business processes. The prerequisite for process centering is a pervasive process orientation throughout the enterprise which, in turn, is based on process thinking. Central to process thinking is the definition of a process as a system put into operation to produce a specific result. The subject process, process feedback, and process quality are classic concepts of process thinking and serve as the foundations for the real-life management of business processes.

Today's business environment of accelerating, unpredictable change demands a change in process design — processes must be more adaptable. While process control is an instrument of stability, it takes on an added dimension in adaptable processes: the assurance of progression from one stable state to another. The management instrument of change and adaptability is process capability.

For an organization to be truly process centered and adaptive, the key element in its success is the people who work in its processes. A business process is not just a repeatable procedure; it is a highly information-intensive, open system in which empowered people continuously adapt the flow of information and work to newly sensed change. Based on the holistic process model, business processes are defined as logical sets of customer-supplier relationships continuously generating commitments which, in turn, drive work activities to produce agreed-upon results.

6 Process Centering: The Basic Approach

6.1 Process Centering as the Prerequisite for Change

Process centering is a business transformation that changes both the function and structure of an organization.[1] It does not mean just improvement, making incremental changes to the structure of what already exists but leaving purpose and function intact. It does not mean just reengineering, either, although it does include the reengineering of individual business processes as its first step.

Before we get into a discussion about how an enterprise can become process centered, we should first understand what process centering is and what its key characteristics are.

6.1.1 Definition of Process Centering

Adapting Hammer's original formulation,[2] we define process centering as a two-dimensional transformation effort that includes:

1. Reengineering of individual business processes through the integration of managing of commitments to customers,[3] workflow, people, and technology for optimal current performance and adaptability
2. Adapting to change through the purposeful realignment of the enterprise around its core business processes

The enterprise is managed as a system in which business processes contribute the most value to corporate performance and prove they are adaptable to change.

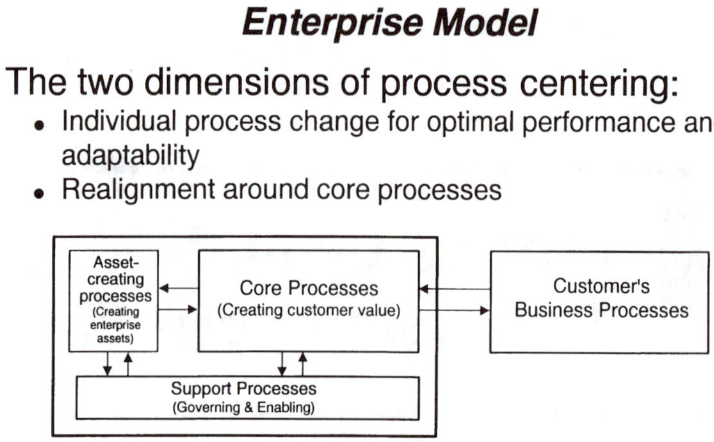

Figure 6.1. The Process-Centered Enterprise

An organization's complex system of business processes is its process architecture. Once established, the implementation of this architecture results in a process-centered enterprise (Figure 6.1). Process centering begins with two assumptions.

The first, as proposed by Ackoff and Haeckel,[4] is that the enterprise is managed as an open system, which means that its parts are in interaction with each other and that, as a whole, the system also interacts with its environment. The second is that the enterprise must be fully capable of responding to change; it can accommodate even the toughest of a whole range of circumstances.

Our practical experience bears out Haeckel's argument. If these assumptions are not clearly understood or accepted in any given situation, the enterprise and its management are not ready to become process centered nor will they be able to deal with the torrent of change that is progressively overwhelming them.

Managing an enterprise as an open system means that process management must focus both on the interactions or relationships (the human factor) as well as on the work activities that make up the business process. Alignment of the enterprise, then, is a state where the purpose of the enterprise, as a whole, is in harmony with the purposes of its parts, that is, its key business processes. This is ensured through the coordination of commitments made to customers for desired outcomes across all interfaces[5] in those processes.

6.1.2 *Commitment Management*

The ability of the enterprise to respond to change is based primarily on process adaptability, which in turn requires a consistent and pervasive management of commitments within each of the enterprise's business processes. Commitment management is not simply the dissemination of a business plan or the distribution of interlocked objectives frozen into an operating plan. Rather, it is the dynamic implementation of customer-supplier relationships throughout the process; it is the existence of customer-supplier dialogues and the conduct of communications at every process interface or hand-off point.

6.1.3 Process Reengineering

The reengineering of individual business processes, with the overriding objective of making them optimally adaptable, will result in effective commitment management and the related communication capability designed into each key business process. To achieve this, each and every member of the organization must understand clearly that process reengineering is *not*:

- Total quality management (TQM), even though it shares a number of fundamental principles with it. Both start with customer needs, recognize the important role that processes play in meeting those needs, and deploy teams for implementation. The key difference between TQM and business process reengineering lies in their respective methodologies and the tools they use. TQM is incremental and, accordingly, its tools are operational; business processes are the vehicles through which an enterprise accomplishes its business strategies and deploys its resources. Also inherent in business process reengineering is the recognition that the deployment of people and information technology is an integral part of the methodology — this is not fundamental to TQM.
- Automation or information reengineering of existing processes through information technology, as they simply provide more powerful ways of doing the wrong things faster and more efficiently.
- Streamlining or downsizing, because these usually mean doing less with less; they are the management methods of contraction. Process reengineering, on the other hand, means doing more with the same or less; it is the management approach to expansion and growth.
- Eliminating the overhead, because overhead and bureaucracy are not the real issue. They are only symptoms of the underlying fundamental problem — namely, fragmentation of work processes in a functionally organized company. In fact, bureaucracy has been the answer, albeit an increasingly inefficient one, to the coordination problems of corporate management of the last 100 years. Overhead and bureaucracy cannot be eliminated without causing disruption, unless the fragmented processes are reintegrated so that the overhead of separately functioning expediters, controllers, supervisors, auditors, and other staff functions are no longer needed for coordination and control.

Traditional business process reengineering is about the reintegration of work and the elimination of process fragmentation. The new process reengineering approach proposed here requires that adaptability become the central attribute of business processes. This, then, is a prerequisite for enterprise-level responsiveness to unpredictable change.

6.1.4 Organizational Adaptability

In today's change-driven business environment, it is organizational adaptability that needs to be achieved, the scope and speed of which exceed that offered by the competition. In this environment, work and, therefore, processes are not rigid; they are naturally adaptable to prevailing requirements and conditions. Organizational rigidity comes from a management philosophy that gives high priority to stability, along with the supporting bureaucratic structures. The traditional command-and-control mentality of management sees change as an unwelcome variation or, at best, an occasional exception to the rule.

When a system seeks to establish equilibrium and stability through constraints on change, it also constrains creativity and adaptability, thus creating conditions that will eventually threaten its survival. Process fragmentation is one such common constraint in the functionally organized corporation. The notion that it was difficult to make more than incremental improvements to processes was a self-fulfilling prophecy promoted during the stable and relatively predictable business environment of the 1960s and 1970s, but organizations that limited their own response to change have faced competitive pressures that have resulted in takeover, significant downsizing, being "spun off," or going out of business.

Conseco, an insurance company that has received much attention on Wall Street, was the target of much speculation in its early days. It had designed a new paradigm for processing and servicing policies. Combining the latest technology and a streamlined operation that allowed fewer people to handle much larger client caseloads, it began acquiring other insurance companies. It basically wanted the insurance policies, not the administrative organizations. The newly acquired policies were entered into the Conseco system to maintain efficiency. The old company's name, in most instances, was left intact so that the customers would not be confused by the change. By the time Conseco was finished with the conversion, only a few employees from each acquired company were needed. For the policyholder, the operation was for the most part seamless. As long as the customer was being taken care of, that was all that anyone worried about. What they did, however, was to challenge conventional thinking.

6.1.4.1 Incremental Change vs. Process Centering

Few instances of successful enterprise transformation have been reported in almost four decades of change management. There exists a host of literature on the successful implementation of well-planned incremental changes, managed through a series of improvement projects, but every time a new level of performance had been established, great effort was made to ensure that control was reestablished so the stability of operations continued.

Nevertheless, incrementally controlled change management and project-by-project improvement themselves are inadequate to respond to the onslaught of massive, accelerating, and unpredictable change in the emerging business environment of the 21st century. The only rational option is transformation of organizations into process-centered enterprises.

Given this logical and almost intuitively obvious conclusion, why is it that most organizations have found it difficult to determine what their key processes would be and what managerial approach and levels of investment were appropriate to achieve a process-centered state? There are two reasons: the narrow, workflow-based, procedural definition of business processes and the inability to determine the true value contribution of the process in question.

6.1.4.2 The Limitations of the Input-Process-Output (IPO) Model

The workflow-based (input-process-output) definition of business processes has been used in most process improvement and reengineering projects over the past several decades. Workflow, however, is only one kind of a process model — best applicable to material flow and manufacturing production processes.

This definition excludes many processes that do not have definable inputs or outputs, or discrete tasks that follow and depend on one another in a predetermined sequence as they pass interim work products from one another. Some of these excluded processes, such as those of the overall governance of an enterprise, its managerial succession, business planning, litigation, acquisitions and divestitures, and many other professional service processes, have as much or more influence on an organization's future as the manufacturing and customer service processes, the usual targets of process improvement.

6.1.4.3 The Value Contribution of Key Business Processes

Studies[6,7] show that, in the first decade and a half of business process reengineering, most corporations used this approach to achieve one or more of these three prime performance objectives:

1. Cost reduction
2. Quality improvement
3. Cycle time improvement

Business processes selected for reengineering were predominantly the core processes which, in some way, directly affect the firm's customers: manufacturing, distribution, customer service, invoicing, and order entry. In the studies, only a small percentage of the companies indicated that their selection was based on some value-based judgment. Even though effective business processes are expected to produce added value, traditional reengineering methodologies so far have not offered any consistent or viable approach to value assessment, because these methodologies concern themselves primarily with improving process performance in current terms of effectiveness and efficiency.

Why is value so important? It is well known that each business process differs in its contribution to an enterprise's overall success and future competitiveness. Key business processes are those which return more economic value than they absorb and contribute the most to the organization's overall ability to successfully respond to change. Aligning a corporation around its key processes means that investing in those processes is more likely to help the enterprise thrive financially than investing in processes which drain the company's resources. This logic leads us to the conclusion that investing in the adaptability of key processes is critical to successful process centering.

6.2 Process Performance and Adaptability

Process performance depends both on how the process has been designed as well as how it is operated.

6.2.1 Definitions

In this book, the term "performance" is used both as an attribute of the process output itself and as a general process characteristic. The process performance measurements that are of prime interest to management are the quality of the work product (process effectiveness), resource consumption or cost (process efficiency), and cycle time — a special type of efficiency measurement. Effectiveness and efficiency are indicative of process performance in *current terms.* Of course, each of these performance categories can be further broken down for ease of measurement and corrective action.

Of the two levels of adaptability identified by Haeckel,[8] our focus here is on adaptability not as a performance measurement but as an inherent characteristic of a business process. Adaptability is defined as the ability to achieve improved levels of performance and to maintain effectiveness and efficiency in the face of change over time. Process adaptability is achieved by rapidly reconfiguring process capability so that the process performance will satisfy new requirements. It is the key to responding to unexpected change; therefore, it is indicative of process performance in *future terms.*

6.2.2 Adaptive Loops in Processes

The key to process adaptability is how rapidly people in their various roles and responsibilities react during the process cycle. The *adaptive loop* is a basic mechanism for the operational management of business processes (Figure 5.2) in that it helps people react to changes sensed through measurement or other forms of observation.

As we will see later, it can be easily integrated into the process management approach built around the holistic process model. There are four steps in the adaptive feedback loop:[9]

1. *Step One* — the correct and timely sensing of change representing new requirements or opportunities
2. *Step Two* — the rapid interpretation of change information with its technical and economic implications
3. *Step Three* — the instantaneous decision about changes required to the existing commitment structure as well as the re-negotiation of current commitments, creation of new ones, and what actions to take
4. *Step Four* — the implementation of actions (work activities) in the process to deliver on the new commitments

Business processes are the producers of business results. All work is done in those processes where the working assets of the organization are concentrated: they are the "muscles" of the enterprise. Just as muscles help humans respond to any number of anticipated or unexpected stimuli by propelling them out of danger or moving them into a new position to deal with change, business processes are the organization's key instrument for responding to change. Like muscles, business processes must be toned to be highly responsive to the demands placed upon them, and, just like muscles, they are activated by signals communicating such things as requirements, opportunities, and commitments. Properly coordinated by an integrated set of commitments translated into measurable process goals, business processes will produce a harmony of value creation across the organization — so pervasive in its effect that the boundaries of contributing individual processes no longer seem to be relevant (Figure 6.2).

Business processes, however, have one distinct advantage over human musculature. They can be reorganized or even replaced; the flow of work and information within a process can be rearranged (reengineered) to achieve breakthrough-level improvements in effectiveness, efficiency, cycle time, or a number of other attributes. If properly designed, processes can learn through the network of learning loops positioned at critical hand-off points within the process and communicating across strategic interfaces with other processes, like humans who play in a symphony orchestra.

6.2.3 The Superiority of Process Centering

The superiority of business processes for managing the flow of information and work in the face of change, then, is well established. Reasons for this include:

- Market power has shifted from suppliers to customers; processes are customer oriented and outcome driven, which is the only basis for establishing a responsive set of commitments and the manageable delivery of results.

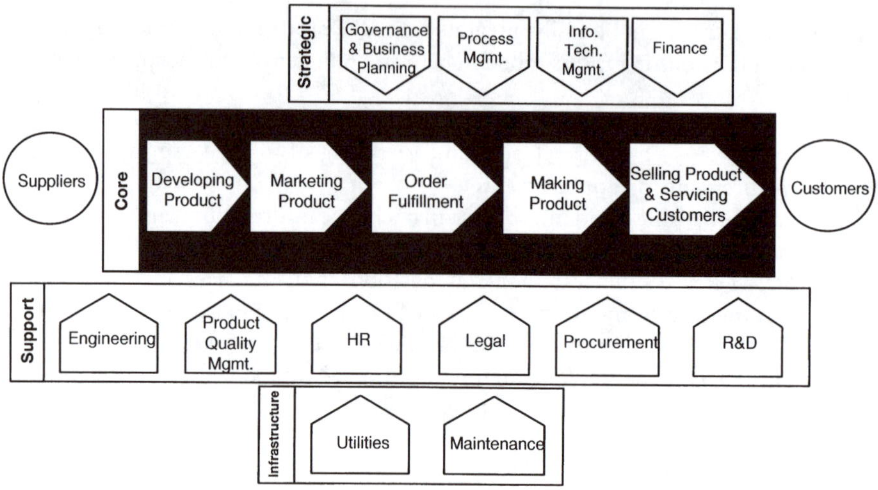

Figure 6.2. Aligned Process-Centered Enterprise Model

- Collaborative relationships are becoming more important; processes translate strategies and plans into chains of activities that actually implement the long-range direction of the business.

Processes are the natural vehicles for system execution, as they bring the physical infrastructure and various systems to life and determine the actual flow of resources — including information and learning — throughout the organization.

6.2.4 Commitment Coordination and Process Alignment

While total quality management and process reengineering have tended to pre-define processes as constructs which their tools can handle, the sense-and-respond approach to corporate governance first proposed by Haeckel[10] is built around the concept of commitment coordination and management through accountability. Communicating and managing commitments and the resulting work allow us to do two things that traditional process management had difficulty doing. It allows us to manage interactions — a chain of customer/supplier relationships, which in turn generate the network of commitments and related communications that drive work activities within the process. This same set of interrelated commitments is also used to purposefully align the enterprise around its adaptable processes to make it highly responsive to unpredictable change.

As we have seen, the poor use of resources in hierarchically managed organizations occurs because a few natural processes, such as product development, production, and distribution, have been broken down into lots of small tasks that people in the functional silos are to do. This fragmented work, then, has to be coordinated and

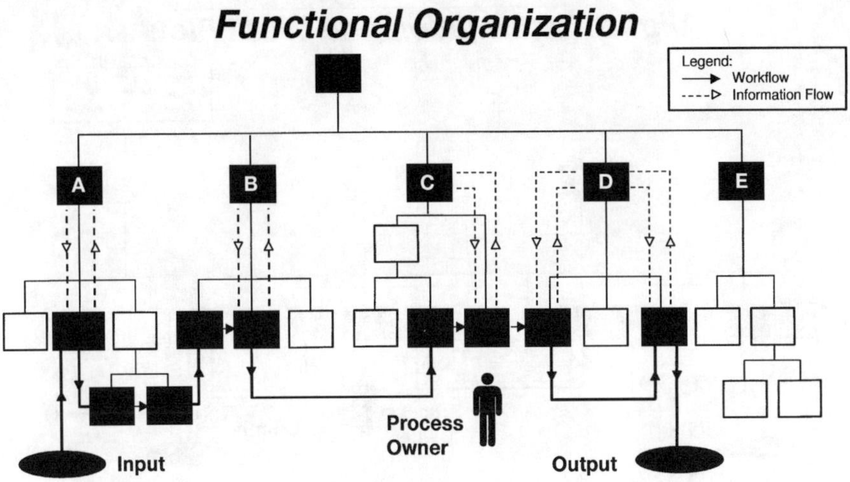

Figure 6.3. Business Process Flow Structure

controlled to produce a single product; it is the supervisors, managers, and staff who are assigned to coordinate the people who do the real production work. This is one major source of product cost.

Such organizations also display an appalling mismatch between the information flow about commitments and the actual flow of work. In other words, the flow of commitments and related communications does not flow side-by-side with the work but takes a totally different direction up and down the management hierarchy. In large, functionally organized corporations, the flow of commitment-related communications is practically prohibited among peer-level employees across functional lines — unless that communication takes place among designated managers and executives at higher levels of the silos and follows the right protocol (Figure 6.3).

6.2.5 What Needs To Be Done

The only logical conclusion is that companies must realign, within each business process, the flow of commitments and related communications side-by-side with the flow of work (Figure 6.4). This simple concept has profound implications in most organizations. Organizations have struggled with communication between work shifts for years. Trinity College's Accommodations Office in Dublin, Ireland, seems to have licked that problem with a few carefully aligned processes. Because their guests do not have telephones in their rooms, voice mail, or mailboxes, the staff has taken on the task of being their guests' eyes and ears. They are all very attuned to taking thorough messages in addition to seeing that they are delivered in a timely manner. Everything is carefully recorded. Shifts overlap slightly. Everyone reads everything at the shift change and asks questions. Everyone considers the lists very

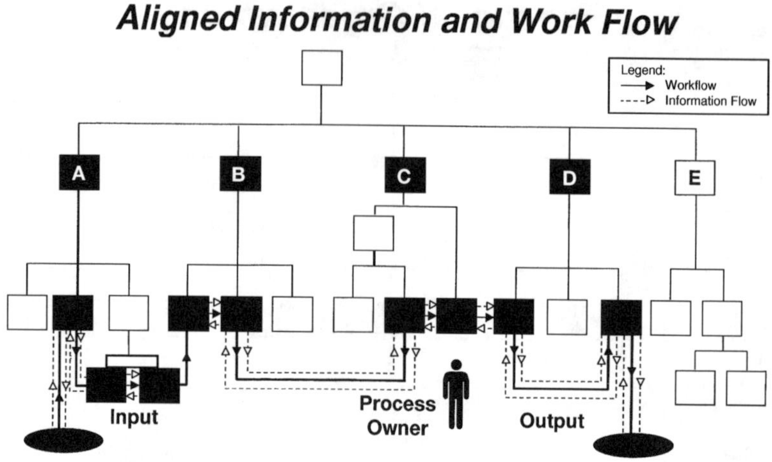

Figure 6.4. Business Process Structure

important and works hard to see that items on them are taken care of — a textbook example of aligning the flow of work and information side by side. These processes were designed by the staff, so each person feels a strong sense of responsibility to communicate quickly and responsibly.

It is not only that work itself has to be reorganized along process lines. What is critical is to realign the flow of communications representing requirements, changes, and commitments with the workflow. This realignment will ensure maximum effectiveness in responding to known customer requirements as well as create the ability to adapt to unexpected change. As we have stated earlier, the process model that incorporates such alignment of commitments and workflow is the holistic process model.

6.2.5.1 The Role of Commitments

The adaptability of business processes is largely determined by the state of commitments that drive the work activities within the process. It is crucial that management be able to rapidly change those commitments that no longer satisfy newly defined requirements and communicate these changes throughout the business process so that the relevant work activities and related resources are deployed to meet all commitments — previous and new. The most important and change-sensitive component of any business process is the people who do the job: process owners and managers who make decisions and agree on mutual commitments, and knowledge professionals and task workers who carry out those commitments by doing the work to deliver on them.

6.2.5.2 The Role of Empowerment

As Haeckel argues,[11] it takes empowered, accountable people, not a stack of procedures, to deal with the unpredictable. Organizations that micro-manage their employees through bureaucratic procedures are doomed long term. To make things worse, oftentimes the procedure becomes more important than getting the work done and producing results.

Organizations trying to adapt to unexpected demands must ensure that the correct people are empowered to deal with unforeseen situations — and then are held accountable for the outcome.[12] These people must interpret the identified change and then initiate a network of commitments among other empowered people within the process to enable the best possible response. In order to do so, they must understand what is required of them and what they have to commit to; they must be empowered and able to renegotiate commitments if they are found to be impossible to meet in the face of change; they must be given the means (knowledge, skills, core competencies, and tools) to deliver on their commitments; and, most importantly, they must be motivated to do all this in a professional, productive, and quality fashion. Such motivation cannot come from campaigns and short-lived initiatives. It must be part of a mental set — an enterprise culture of accountability which, even in the absence of outside motivators, can still be self-motivating and self-fulfilling.

One of the most difficult challenges of product development is the introduction of design changes after the development process has entered the "frozen zone." At this point, the product specifications are finalized ("frozen"), the product is committed to be built against them, and the process owner (or product manager) becomes accountable for delivery on the commitment. Procedural rules of the development process simply do not allow any further changes to be introduced into the design. However, for persuasive reasons, the person accountable for the committed design may accept further requests for design changes, as he or she is empowered to do so — and, accordingly, renegotiate the existing commitment to meet the new requirement.

6.3 Summary

Process centering of enterprises must be based on a very fundamental assumption about what the organization is to do. What it does is done in its business processes. The organization hoping to be competitive in the 21st century must be designed to adapt rapidly to accelerating and unpredictable change in customer requirements, technology, and the environment. To be capable of doing this, enterprises must be process centered — aligned around their adaptable core business processes. Process adaptability, in turn, requires a consistent and pervasive management of commitments within each of the enterprise's business processes.

Business processes can be reengineered for adaptability; the underlying condition for doing so is to take the holistic view of the process, which includes both the management of commitments (the human factor) as well as the work activities that make up their execution (the procedural part). The key to process adaptability is how rapidly people in their various roles and responsibilities react in the process cycle. The adaptive loop is a basic mechanism for the operational management of business processes in that it helps people react to changes sensed through measurement or other forms of observation. Organizations trying to adapt to unexpected demands must ensure that the proper people are empowered to deal with unforeseen situations — and then are held accountable for the outcome.

7 Process Centering: The Response to Change

One of the continuing messages of this book is that earlier attempts to handle accelerating change were just initial steps on the road to creating a process-centered organization. Organizations that have been through those attempts have already learned some valuable lessons. To be truly process centered, though, requires efforts well beyond what has already been tried.

7.1 Response to Change

In the final analysis, downsizing and other traditional management practices such as total quality management and process reengineering have made companies more efficient with a resulting over-capacity in many cases. But enterprises presently poised for the resumption of growth have also begun to "upsize" their workforces again; IBM, General Motors, Hughes Electronics, Daimler-Benz, and AT&T, all major downsizers during the mid-1990s, have begun to recruit again and add new increments of people to their payrolls.

7.1.1 Upsizing and Growth

As companies contemplate opportunities to resume growth, a few key questions must be asked:

1. What kind of growth should companies seek (whether they were downsized or not) as they move into the era of upsizing?
2. Will the upsizing just be building back old capacity and past employment levels? Or will it be growth into new markets and new lines of products and

services, perhaps even entirely new businesses? And what growth strategies should be applied?
3. What type of skills will employees require?
4. Is it clear that the company is fit and ready for growth in an expanding, but quickly and unpredictably changing world economy? Or will it succumb helplessly, as have scores of bankrupt companies, millions of newly unemployed, and hundreds of thousands of impoverished investors in Eastern and Southern Asia?

As we will show below, the growth strategy of choice is that which results in enterprise development,[1] along with growth in size, capacity, volume, and profit — because this strategy is a corollary to investment in process adaptability and process centering.

7.1.2 The Nature of Change

As we have seen earlier in the book, change can be predictable and incremental — or unpredictable, sudden, and radical. Its nature and sources, however, have evolved over time to the point where, today, the acceleration and unpredictability of change have led to an essentially turbulent economic, technological, and business environment.

7.1.2.1 Continuous Change

Until the 1960s (the decade that also saw the massive introduction of the digital computer into industrial and business use), economic and technological change was, by and large, continuous. Response strategies first attempted to control change (change management), so that the forecast-based, planned approach to business management could continue undisturbed. Later, management practices more responsive to continuous change were introduced: continuous improvement and organizational effectiveness were the methods of the 1970s and 1980s.

Partially due to the rapid development of information technology, not only did continuous change accelerate, but it also became punctuated with occasional, sudden, accelerating, and unpredictable change.[2] Process reengineering was added to the arsenal of management technology as a tool better suited for dealing with this occasional situation on a more overarching scale.

There are two different modes in which any given organization can find itself as it embarks on its response strategy to deal with continuous change: defense or offense. In a defensive mode, management is trying to maintain and possibly improve performance and related operations. Here the main concerns driving corporate response include:

- Protecting profitability
- Maintaining growth
- Safeguarding competitive position
- Maintaining or improving customer satisfaction and quality

In an offensive mode, the intent is more that of prevention and interception, or taking advantage of an opportunity presenting itself. Here the main factors behind the organization's response include:

- Accelerating or slowing future growth
- Preventing potential loss of organizational control
- Meeting or leapfrogging expected technological change
- Anticipating changes in the political, legislative, or regulatory environment

7.1.2.2 Accelerating and Unpredictable Change

By the early 1990s, it became clear that the continuing explosion of recorded information and knowledge, along with increasingly easy and affordable network access to them, was creating a new source of extremely rapid, accelerated, and unpredictable change. Businesses have become capable of creating and manipulating massive amounts of information at incredible electronic speeds. This information often represents tangible items, such as products, inventories, or other capital assets. Sudden and unexpected changes in the real world can appear that need to be managed. A good example of this is the phenomenon known as flying by wire, described by Haeckel and Nolan:[3]

> *In aviation, flying by wire is the method of piloting an airplane developed in response to the changes (especially speed) introduced by jet-engine technology in the 1950s. Flying by wire means using information technology — communications and computer systems — to enhance the pilot's ability to sense, interpret, and react to the rapidly and unpredictably changing environment around him. Heads-up displays (computer-generated pictures projected onto the pilot's helmet visor) present selected abstractions of crucial environmental factors such as oncoming aircraft or missiles, targets, and the like. Information technology aids the pilot in evaluating response alternatives. The pilot's decision (say, to take evasive action or fire a missile) is then translated into a computer command which, in turn, is converted into thousands of detailed instructions that orchestrate and define the behavior of the plane in real time. The essence of all this is that when pilots fly by wire, they are flying informational representations of airplanes.* (From Haeckel, S. and Nolan, R., Managing by wire, *Harvard Business Review,* Sept./Oct., 1993, p. 122. With permission.)

Electronic commerce — also known as *e*-commerce — is essentially the trading in informational representations of goods and services. This includes marketing, buying, and selling products, from cars to computers, over the Internet. The physical product is eventually delivered, but only after all business transactions are completed electronically.

Another well known, but less understood, new area of the information economy is *e*-trading, the multi-trillion dollar worldwide business of financial derivatives trading. Derivatives as products are created, bought, sold, and eventually retired in cyberspace: no one physically sees the actual derivative product. People working in this environment are those who not only feel comfortable with, but are highly effective in, manipulating symbols or images as if they were tangible reality.

But the real business transformation made possible by the networked economy is *e*-business: the true information partnership of an enterprise with its suppliers and customers — to achieve the "fully integrated value chain."

What we are experiencing today, then, is the emergence of an extremely information-intensive business environment in which continuous, accelerating, and unpredictable change has become the norm. Responses to such external change at the enterprise level must occur not only in the structure of processes and organizations but also in the function of the enterprise defined by its purpose and image as well. This is because continuing unpredictability dictates a fundamental internal change in the business mission. This type of change, known as transformation, no longer concerns itself with just improving (or reforming) what exists, but rather with taking the enterprise into totally new territory.

7.2 Response Characteristics

Generally speaking, the characteristics necessary for appropriate responsiveness to change are information intensity and process robustness. The two are related, because robust processes depend on the flow of good and timely information to reconfigure and adapt themselves when needed.

7.2.1 Information Intensity and Process Adaptability

Given the unpredictability of change, individual business processes, as well as the enterprise as a whole, must become highly information intensive. Process adaptability becomes virtually synonymous with information intensity. The consequence of this is that an information-intensive process is capable of change as new information becomes incorporated into it.[4] The process is no longer static or procedural; it is dynamic and flexible because it is capable of adapting the way it performs work in response to newly received information inputs.

Information intensity at the process level means that satisfactory products and services result from collaborative interactions between customer and supplier:[5]

- The customer supplies information about needs, expected value, and conditions of satisfaction against which the supplier is to deliver.
- The supplier provides information about capabilities and performance expectations so that this dialogue leads to greater value creation.

For example, consider the customer who needs a service that can deliver packages overnight. The supplier states its capabilities by asking the customer to get the package to the designated acceptance center by 9:30 p.m. If this condition is met, the supplier's commitment is that the package will be delivered the next day. The business is driven by customer requirements, not by what is in inventory or what has been done previously.

Predictions are that in the not too distant future, car buyers will have two choices: a custom-made vehicle or a used vehicle. Automakers will no longer make cars without knowing who the buyer is. Consumers will be able to order their cars directly over the Internet, and the new car dealers and used car specialists will be eliminated from the ordering process. In addition, the new car will be delivered directly to the buyer's home if he or she does not want to pick it up at a delivery location. The entire order/delivery cycle will take less than a week. The changeover has already begun, with the automakers combining and closing dealerships worldwide. With this configuration of doing business, what everyone will have to sell is excellent service.

As the information intensity of a process increases, a number of related developments occur:

- *Product life cycles decrease.* As offerings become more information rich, it becomes desirable, or even necessary, to change (upgrade, replace) them more frequently. This development leads to the growth of productivity. The impressive growth of Nokia, the Finnish cellular telephone manufacturer, is a good example of this phenomenon. As the information-processing capacity of cellular phones increases, so does the number of application opportunities for their use. The result is the increasingly frequent introduction of new models with an expanding number of features and capabilities.
- *Process cycle time decreases.* With the need for increased frequency and number of offerings, the length of each transaction (process cycle) is, by definition, reduced through the application of suitable process reengineering methods. Cycle time reduction can also lead to growth in volume and revenue. Today's software industry is among the leaders in mastering cycle time reduction in order to accelerate the release of software version changes and updates to their customers. Major operating system releases now routinely occur within 12 to 15 months, as witnessed by the successive shipments of Windows 6, Windows 95, and Windows 98 to the marketplace by Microsoft. Version updates of major software packages now take but a few months. Users of Norton antivirus software and Microsoft's Office 97 can download enhancements about every month now.
- *The number of transactions (sales, service experiences) with the same customer is maximized.* Each transaction becomes more important and profitable through providing a more diverse collection of offerings. A side effect of this is the steady growth of the number of loyal customers. For example, Kinko's has combined a number of services under one roof for the small-business owner or professional: faxing, printing, mailing, computer design, stationery store supplies, graphics presentation materials, and much more.
- *Capability self-enhancement* is another desirable response characteristic at the process level. During the customer-supplier exchange of information, when

an inability to respond to a new requirement is recognized, the process must invent, enhance, acquire, or create a new capability in order to meet the new, higher level expectation. Adaptability, in essence, becomes an *ad hoc* invented response to unexpected change. The invention of a new tool can result in significant capability enhancements. An example of this is the digital computer that essentially accelerated, by orders of magnitude, the execution of already existing business processes.

An everyday approach to the acquisition of new capabilities is outsourcing. Many companies have decided to transfer work, previously done by their own employees, to outside suppliers or vendors. Cases in point include the range of activities from information services to cafeteria operations.

7.2.2 Process Robustness

Traditional work processes were first designed in the manufacturing industry to provide mass production for a mass market. All inputs were predefined and identical; accordingly, they were also handled identically. Acting upon identical inputs, process operations carried out identical sequences of steps to produce uniform and consistently similar outputs. In this world, variation in process flow or, worse, process output was to be guarded against; when it occurred, it had to be controlled and eliminated. Variations used to be dealt with harshly in most organizations.

The theory and practice of quality control were developed around the concept of minimizing or eliminating variation. Variation in output was correctly interpreted as a process defect or outright failure, because processes, by definition, were supposed to be stable and repeatable. In today's world of diverse markets and changing requirements, this procedural logic is obsolete and no longer tenable. How to deal with variation is the new mantra of many organizations, because ability to do so creates new markets and new business. Instead of mass uniform production, mass customization is the wave of the future.

7.2.2.1 Multi-Variant Processes

To meet the demands and expectations of today's business environment, we need the flexibility of executing any number of variants of the same process, each one responding to the changing requirements of the same or different customers, situations, or inputs. Still, these variants of the same process must have the same performance characteristics and economies of scale that would result from predictable mass production.

Flexible manufacturing processes are based on this concept. They can be executed in several variants, depending on what product type is to be produced. For example, in the auto industry, one variant of the same assembly process will produce a sedan, and another version, a van — using different parts and a number of different work steps, but being essentially drawing upon the same basic process.

Another example of this approach is Levi's mass customization of jeans. Thanks to a sophisticated information technology that allows the gathering and storing of vast amounts of customer measurement data, the same process used to produce a small number of sizes is now used to produce jeans that fit an individual customer's body perfectly.

Process designers have learned the hard way that to create a one-size-fits-all procedure to accommodate all eventualities is not only very complex but simply unaffordable to do. Such designs have to incorporate special rules, situations, and a wide range of possibilities that can be anticipated or even thought of as exceptions to handle the other exceptions. A multi-variant, self-adjusting process, by contrast, is relatively simple and straightforward from a design standpoint because any given variant needs to handle only the situation for which it is intended. There are no special situations or exceptions in the process flow itself — they are handled by the commitment management part of the process, whose content is all information and communication.

Using the variants of a process, pharmaceutical companies have discovered entire new markets for drugs already in development or on the market: a drug developed for one specific disease may be found to be more effective for another. The most recent example for this is Viagra, originally developed by Pfizer for treating certain cardiovascular diseases such as angina. However, in the clinical test phase, the drug did not appear particularly effective on patients with coronary heart disease. At this point, no obvious pharmaceutical target remained, and the firm considered terminating the development project. The very meticulously kept clinical test records, however, showed a very definite pattern of side effects, indicating to developers that the drug would show much greater efficacy in the treatment of certain types of pelvic psychogenic diseases — particularly in men. This, then, became the new target area for the drug. The rest is history: the drug that showed only mediocre results for treating one disease has become a true blockbuster in treating another. What happened from a process standpoint was that good scientific observations were conducted through comprehensive information gathering, recording, and knowledge-base building. This made the process highly information intensive. While a rational approach to treating one disease did not succeed, the drug developed for that disease helped researchers better understand the nature of another. The knowledge base related to the workings of the drug grew as the clinical tests progressed. The development process was robust enough to allow rapid change of direction and movement toward hypothesis testing in the new target disease area. Research management at Pfizer was responsive enough to recognize the opportunity and make the right follow-up decisions.

7.2.2.2 Proximity to Competency

Another important responsiveness characteristic of the redesigned process is that work is concentrated and performed where the competency and resources reside — that is, where it makes the most sense from the standpoint of the ability to respond and deliver. In the traditional functional organization, work is organized around specialists, resulting in specialized functions or departments across the organization. A good example for addressing this situation is the reengineering of the traditional purchasing practice in the Central Research Division of Pfizer Corporation — leading to significant productivity growth. Research scientists know how to conduct research, accountants know how to count, and purchasing clerks know how to order supplies. So, when the research department needs new supplies, the purchasing department buys them. Purchasing finds suppliers, negotiates price, places the order, verifies it, inspects the supplies upon receipt, and eventually pays the supplier's invoice. In due course, the scientists at Pfizer get their supplies, unless the approved supplier is out of stock and the purchasing department decides on an alternate source.

This kind of process is expensive as well as time-consuming and wrought with possibilities for error, dissatisfaction, and delay. In any given situation, it involves at least two departments (original requestor of the supply and the purchasing department), plus the overhead associated in both departments with tracking all the information and paperwork until all the pieces of the process have come together. The idea of spending more internally than what the purchased item cost did not sit well with management at Pfizer, so they decided to transfer the responsibility for purchasing supplies onto the customers — in the case of the research department, the scientists, who now buy their own supplies, pencils, printing paper, subscriptions to journals, books, and other low-value research supplies such as chemicals, etc. The customers know from whom to buy and what to pay because the purchasing department, now into strategic procurement planning, has established preferred supplier lists, negotiated preferential prices, and given the accounting department a list of approved suppliers who may be paid directly without undue review and controls.

At Pfizer, operating units, or even individual employees, have authorizations to use a billing card with an established line of credit. At the end of each month, all card transactions are reconciled against the internal general ledger system so that individual department or employee's budgets are charged for the supplies ordered. As a result, customers receive their supplies more quickly, and the corporation spends a fraction of what it used to on internal processing costs. What's more, if requirements change or new suppliers appear who offer a new set of products or other goods, the process is flexible and adaptable enough to accommodate these new arrivals and opportunities without changing any of the basic rules or parameters of the process. Authorization levels can be modified by a single telephone call or e-mail message; new suppliers can be added to the approval list by a simple, straightforward procurement management decision. The overall result is significant growth in productivity and efficiency. Another manifestation of proximity to competency is known as *concurrent engineering* (see Section 10.3.6.4).

7.2.3 The Economics of Increasing Returns

To transform an enterprise into an adaptive open system aligned around its key business processes (and then to manage it so that it can optimally respond to new and unexpected changes) depends on the information intensity of the enterprise itself.

Information intensity of an enterprise leads to different economic characteristics and behavior that are highly relevant to the kind and magnitude of growth the enterprise seeks to attain.

As corporations become more information intensive, they are more likely to become subject to the economics of increasing returns. It has been known for some time that this class of economics prevails in periods of rapid technological change, where a firm's initial success gives it an advantage that can be progressively leveraged into more and more advantage. The larger a firm grows, the greater leverage it will tend to have. Knowledge-based parts of the economy are primarily subject to this theory. Some of the more recently emerged technology leaders are examples of this. It is interesting to speculate which other corporations of today's information age are likely to be the next beneficiaries of increasing returns economics.

The spectacular growth of Microsoft, Intel, and AOL, with their geometrically increasing leverage over their respective markets, seems to provide the empirical validation of increasing returns. On the other hand, it certainly did not apply to IBM at the peak of its performance in the 1980s. While IBM was producing and selling information technology worldwide, it was not exploiting the information available about its customers' use of that technology and, consequently, it was unable to take advantage of the opportunities this presented.

7.3 Response Strategies for Growth

As Ackoff[6] points out, the need for growth comes from the basic logic that in today's quickly changing and competitive global environment only those enterprises able to thrive will be those that can continually increase their productivity. By definition, to increase productivity, the labor content of a unit of output must be simultaneously decreased, which leads to steadily diminishing employment. However, enterprises must maintain some steady level of employment so that they can benefit from human creativity, learning, and knowledge needed, in turn, for the design and implementation of further increases to productivity.

Stable employment and continuing productivity improvement require simultaneous, attendant growth; therefore, in an enterprise defined and managed as an open system, growth must be seen as a means, not an end. The end, as previously mentioned, must be enterprise development: the increasing ability to satisfy known needs and expectations in the economic, informational, and human dimensions.

7.3.1 Processes as Product Offerings

A desirable response characteristic at the enterprise level is the ability to turn processes into product offerings. An example of this is Xerox's ability to manage customers' copying facilities as well as supply their equipment needs, which created a new and lucrative market. When Xerox created its Business Services unit and introduced facility management as a service offering, it turned a set of its existing support processes into a highly sellable and profitable product. Its traditional equipment planning, maintenance, and technical repair processes became the basis for a customized service offering that is generating significant additional revenue for the corporation. Its profitability derives from the fact that the service is built around already existing hardware, equipment, and facilities; hence, no massive new capital investment was needed.

When an existing process is turned into a product offering in a new way, the enterprise generates new revenue from investment it has already made and generates significant economic value added (EVA) in doing so. For instance, EVA occurs when a book publisher sells its publishing process to corporations for producing product and service documentation for their customers. This can also lead to escalating volume and revenue growth opportunities, as product-turned-processes usually represent entries into new applications and markets; otherwise, they would not have been initiated in the first place.

7.3.2 Market Preempting

Another desirable response characteristic at the enterprise level is known as *market preempting*. It is the ability to use processes to capture other industries' customers and enter entirely new markets. Successful preempting demands an appropriately designed and effectively performing process, along with a robust customer knowledge base and timely access to customers. Enterprise image and reputation can also be key to the customers' willingness to switch suppliers and buy the new offering.

When Morgan Stanley decided to experiment with online discount buying, it was one of the first of the large Wall Street brokerage firms to do so. Rather than being faced with mass defections of their clients, they decided to take the initiative and create a new part of the enterprise that is leading the world to a new way of buying and selling stocks within the context of an old, established brokerage firm.

The importance of the adage, "know your customer", in preempting is illustrated by an auto insurer expanding into financial services. The United Services Automobile Association (USAA) has focused on insuring automobile owners in America's armed services. Even though its customers work in a moderately paid and progressively shrinking industry, comprehensive knowledge of the needs of those in military service and their families helped USAA grow consistently since 1988 and successfully branch out from car insurance into other areas of financial management for the same customer set. Another example of preempting includes British Airways, which moved aggressively into the hotel reservation business where it already had an existing sophisticated, computerized worldwide reservation system that assured superior competitive position right at the outset.

As the information explosion helps the boundaries between the various industries dissolve, preemption will be an increasingly attractive change-response strategy. Over the next decade, we can expect the major accounting firms to preempt law firms for certain types of legal services; investment houses are already preempting banks and vice versa, and banks are preempting travel services.

Thus, preempting may turn out to be a double-edged sword. In developing its change-response strategy, the enterprise should look at its own preempting opportunities but also be positioned for the possibility that others will attempt to preempt some of its current customers. As to economic results and growth potential, preempting is in many ways similar to turning processes into products in that it generates new revenues from brand new customers and markets; it also creates new EVA and opens up the potential benefits of increasing returns economics for the enterprise.

7.3.3 Process Investment Strategies for Growth

Because of today's information-intensive business environment and accelerating, unpredictable change, any response strategy for the attainment of growth must be based on continuous but selective investment in the enterprise's key business processes. Investment is always a question of potential risk against potential return. Clearly, not all business processes should be invested in over a given period of time. Many offer solid performance improvements and even significant cost reductions but do not create any significant EVA. Others offer opportunities to create EVA, but no appreciable growth in the information intensity or knowledge base of the enterprise.

A favorite target of early reengineering efforts was the customer service process with the prime objective of reducing customer complaints and shortening the service response cycle. While very often these objectives were achieved and even exceeded, the cost of reengineering, including the automation that went hand-in-hand with it, offset any potential economic value added. Clearly, most of these types of reengineering efforts were suboptimizations at the individual process level.

Establishing the direct link between business processes and capital investment must precede centering the enterprise around its key processes. This is made possible by recognizing that business processes represent assets and liabilities — they are a major part of the company's capital base. Growth, then, becomes the measurable return on investments made into certain business processes over a period of time. With this investment mentality in place, four process investment strategies emerge that, individually or collectively, will provide an optimum path to future growth and development.

7.3.3.1 Strategy One: Customer Franchise Management

Customer franchise management[7] invests primarily in the customer knowledge base of core business processes. The result is a careful, complete, in-depth definition and description of the enterprise's customers and their needs, expectations, preferences, habits, and histories so that products and services can be provided that are clearly superior to those offered by competitors. Companies are investing heavily in market intelligence computer bases. One big discovery has been that, by interviewing employees who interface with customers regularly, the amount of previously unknown data about those customers is almost staggering. Also, in the past, most companies have failed to put their sales data online so that the employees interfacing with the customers can understand the extent of their company's relationship with that customer. These databases have become extraordinarily helpful to all parts of the business because the relationship and commitments with the customer have been abundantly clear. The key to this process has been universal access so that no filtering of information distorts the reality of the situation. Some managers and upper-level executives have felt threatened by this because they no longer have the power they once had. Control of such information has been very well protected by upper-level management in the past. As with all computer databases of such a sensitive nature, companies need to develop and install adequate information protection mechanisms.

In essence, this is a pervasive implementation of the customer-supplier dialogue across all business processes that touch customers, with the prime economic result of growth in profits (dollars) and profitability (margin) and the related flexibility in financing other areas of growth within the company.

7.3.3.2 Strategy Two: Steady Stream of New and Innovative Products

Offering a steady stream of always new and innovative products requires investment in two areas:

1. The core processes that provide added customer value, such as development, manufacturing, distribution, and service
2. Those processes which, by "teaching" other processes, contribute most to the management of learning, innovation, and knowledge within the enterprise (i.e., research, benchmarking, and data management, among others)

One measurement of the success of this strategy is the percentage of annual revenues (or profits) derived from products introduced over a given period of time (for example, in the last five years). Master practitioners of this strategy include 3M, Gillette, Hewlett-Packard, Microsoft, and Rubbermaid.

7.3.3.3 Strategy Three: Share the New Wealth

Increasing the share in new wealth creation is one of the boldest and most imaginative growth objectives the management of an enterprise can set for itself. While EVA measures the ability of a company to produce returns in excess of its cost of capital, it reveals little about the share of the company in new wealth created in an industry or what is known as the relevant domain of that company. A rough measure of the current share of a company in new wealth creation can be determined by comparing the company's current share of total market capitalization in the company's relevant domain with its market share, say, a decade ago.

In 1988, IBM's share of total computer industry (including office automation, where IBM was a major player) market capitalization amounted to 45.9%.[8] By 1997, it had plummeted to just 14.2%, even though at the same time IBM was earning a healthy 17% return on invested capital. It seems that the company's business strategy now does include something more than fueling its ability to produce an economic profit year after year while new start-up competitors are capturing the new wealth generated in its own industry.

To begin focusing on share in new wealth creation, corporate executives must be able to sense the needs and opportunities in their industry and beyond. To this end, investments should be made in sensing earlier and responding more quickly to those processes that are the prime repositories of the company's information intensity, technical intelligence, and market knowledge base. Above all, this calls for investment in the adaptability and self-enhancing capabilities of processes that foster creativity and innovation: research and development, information technology, knowledge management, and human capital development.

7.3.3.4 Strategy Four: Enhancing Information Intensity

This strategy calls for steadily enhancing and improving the information intensity of the enterprise through investment into processes that generate and manage information, and contribute to the growth of the enterprise's knowledge base. This strategy is based on the emerging recognition that information and knowledge have become primary economic resources, gradually replacing capital and energy; the economic behavior of information is vastly different from that of capital, energy, and labor; and the economic value of information can only be measured through use over time.

Information intensity is greatly supported and enhanced by comprehensively designed and functioning software systems. Baan, Oracle, PeopleSoft, SAP, and others have developed integrated software known as enterprise resource planning (ERP) systems. This type of software has become very much in demand in recent years and fueled tremendous growth for these companies. The software basically integrates all the databases in all parts of the organization, thus creating part of the integrated enterprise intelligence mentioned earlier. In manufacturing, it allows people to track what is occurring during every step of the process from order entry to product shipment. It has also enabled companies to communicate with one another through the use of common software and shareable databases. Rather than contacting a salesperson, a company can order more of a product directly via specially configured computer order lines; the system calculates how quickly the order can be filled. It then automatically revises the production schedules to reflect the new orders. ERP systems are highly scaleable; therefore, people can manage the growth without also having to manage a myriad of new detail which used to be entered manually and subject to a lot of error. Installation of ERP software is a massive undertaking that is labor intensive and expensive up front, but it enables a set of interconnected processes to be far more capable of dealing with change.

7.4 Summary

In our age, the continuing ability to respond to accelerating, unpredictable change also offers new opportunities and choices for attaining and then enhancing business growth. The two desirable process-response characteristics are information intensity and process robustness. These process characteristics can be enhanced by appropriately chosen response strategies at the enterprise level, such as offering processes as products, and market pre-empting. Because of today's information-intensive business environment and accelerating, unpredictable change, any response strategy for the attainment of growth must be based on continuous but selective investment into the enterprise's key business processes. With this objective in mind, several process investment strategies emerge:

- Customer franchise management
- New and innovative products
- Sharing the new wealth
- Enhancing information intensity

They contribute not only to effective response, but also to growth combined with enterprise development — the kind needed in an organization defined and managed as an open system.

8 Process Centering: Role of the Individual

One of the basic arguments of this book is that it is empowered, accountable people working as individual professionals or as members of a specific process team who make the difference in making individual business processes adaptable and, as a consequence, the enterprise itself responsive to change. This chapter reinforces the critical part each individual plays in process centering the organization.

Whenever change occurs, most people are not resistant to it as long as management clearly spells out what can or needs to be done by each individual — in other words, what each person is empowered to do. One of the biggest problems with reengineering has been that all the planning is exceptionally well done, as planning is the easy part, but the execution of the plan has been done poorly. This is particularly true when unexpected changes occur which necessitate deviation from the plan. The execution of any plan in the face of change depends on each individual within the process being very clear about his or her roles and responsibilities. Those responsibilities include what is expected — awareness of what is actually happening, and who is empowered to make needed decisions about how to respond to the changed situation.[1]

Failure to recognize this may doom the reengineering effort and, most importantly, can result in business failure. There have been a number of lawsuits in the past few years concerning the failure of businesses after their reengineering efforts rendered them incapable of carrying on their day-to-day business as they reconfigured their internal processes.

The key to successful change is the people. They will succeed only under these three conditions:

1. They must be *empowered*[2] (they are given the authority and responsibility for their jobs).
2. They are *committed* (they know their roles and capabilities and believe the change is critical to the success of the organization).
3. They can be held *accountable* (they understand and accept the consequences of having made their commitment).

Performance successes or failures within well-defined parameters are then their accepted responsibility.

8.1 Process People

Traditional, hierarchical business structures have operated in a context of survival and security, and they have been made to work through authoritarian command and control. The purpose of the military is not only to survive in the face of the enemy, but to prevail upon them and thereby ensure the level of security required. It is a classic example of a hierarchical structure that could not be made to work effectively through anything but command and control.

The currently emerging process-centered structures operate in the context of conscious choice — which, in turn, is created through open communication and commitments freely negotiated and agreed upon by empowered and accountable people. Accordingly, process-centered structures are made to work through empowerment of employees and commitment to relationships and agreed-to outcomes.

Satisfactory results, continuous improvement, and the ability to respond to unexpected change can be attained only by building a framework of commitments with everyone involved. A business process can be viewed as the organized combination of commitments, communications, and work to produce a desired outcome. In such a process, the following occurs:

- Resource-consuming assets (which include humans) carry out work activities that produce outcomes such as products, results, or services; the flow of work is highly procedural and predetermined.
- The reasons for the work done in the process (process mission, goals, and performance objectives) are the result of human planning. Those reasons are determined usually through a series of formal dialogues described by communication protocols and guided by a set of agreed-upon commitments made to customers and in force at any given time.
- The totality of these commitments, in turn, represents the commitment state of the process at that same time; it is susceptible to change, highly variable, and capable of improvement.

To further develop these concepts, a few definitions and examples will be helpful.

8.1.1 Empowerment, Commitment, and Accountability

Because each individual needs to understand what is expected of him or her in the process-centered enterprise, the old passive "do as I say" approach to doing work will not suffice. Employees will need to be selected for their expertise and ability to adapt to a changing environment, trained and empowered, and then set loose to do their jobs. Evaluation of performance will be based on outcomes: how the individual meets his or her commitments, not on how he or she follows rules and procedures.

> *A commitment is defined as a negotiated agreement between two parties to produce and accept a defined outcome, subject to agreed-upon conditions of satisfaction.*

Because people must make commitments, they need to be empowered to do so. If a customer service representative in Union Camp's Fine Paper Division's order-fulfillment process says a product shipment will be delivered on a certain date, then the entire organization works to honor that commitment.

Empowering means providing the skill, competencies, information, methods, and capabilities required to reach agreements and to make and fulfill commitments. Being empowered is a state; commitments are made possible by empowerment. If a 500-ton shipment of inkjet printing paper is delivered to an incorrect location, and the customer needs it within 24 hours, the transportation people should be able to make the decision to reroute the shipment (even if requires a different carrier and higher expense) without having to obtain a raft of authorization signatures.

Following Haeckel's approach, we define accountability as the *personal acceptance of the consequences of making a commitment.*[3] Only people can be held accountable; organizations, systems, processes, or machines cannot. Accountability specifies who is committed to produce what outcome for whom; it arises exclusively from outcomes agreed upon and committed to among empowered people. Accountability is the glue that holds customer-supplier relationships together; the supplier is accountable for producing the defined outcome; the customer is accountable for accepting it if it is produced within the conditions specified. As a result, there is no such thing as generic accountability, such as someone being held accountable for marketing, or sales, or manufacturing, or quality, and so on. Accountability must always be linked to a specific outcome.

Outcomes produced by a business process flow from the sum total of process capabilities that exist at a given time. A capability is defined as the inherently uniform set of outcomes that can be produced by resource-consuming process assets performing specific work activities. If one is accountable for a given outcome, one should be provided with the authority and empowerment to mobilize and apply the capabilities needed to produce and deliver the outcome committed.

The capabilities of a business process are usually represented by its subprocesses, whose combined output contributes to the overall outcome produced. The linkage of these basic concepts is illustrated in Figure 8.1. In a business process, the management of commitments and the related formal communications are indispensable in

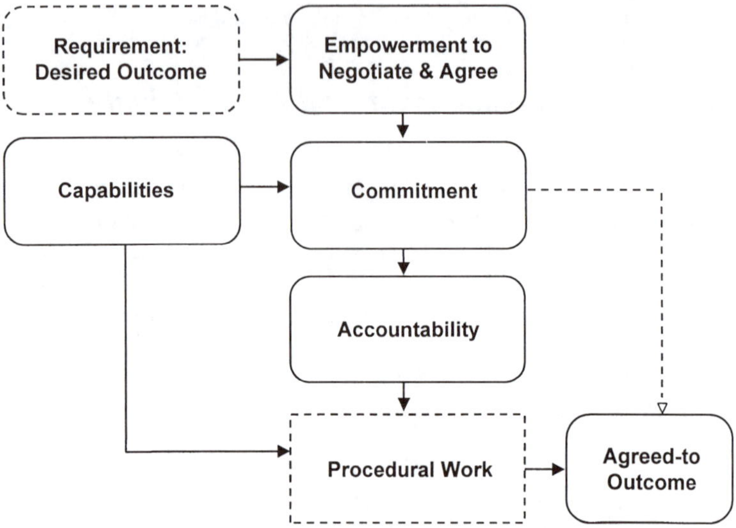

Figure 8.1. Empowerment, Commitment, and Accountability

determining the initial set of capabilities (actions, procedures, and resources) that make up the process. Also, and more importantly, unpredictable changes in the environment will generate new requirements that, in turn, will make it necessary to identify and agree upon appropriate process changes (e.g., improved levels of capabilities).

8.1.2 The Process Professional

All work performed in a business process and, by extension, in the process-centered enterprise is process work. The procedural part of this work is performed by the resource-consuming assets of the process: people, materials, energy, and equipment using appropriate information. Its result is the desired process output or outcome. Some work, however, does not contribute directly to the process output; rather, it is needed to negotiate, make decisions, and communicate commitments. This type of process work is highly information intensive and comprises primarily a series of disciplined and formal person-to-person, customer-supplier communications. Ultimately, it drives the procedural work in the process.

The people performing these two types of process work may come from a variety of disciplines and occupations: assembly workers, clerks, security officers, secretaries, administrators, engineers, lawyers, accountants, managers, and so on. What counts from a process viewpoint is that they bring needed and unique sets of skills and competencies, without which the process could not be performed as required. Because of their equally important contribution to the process, they are all considered *process professionals.*

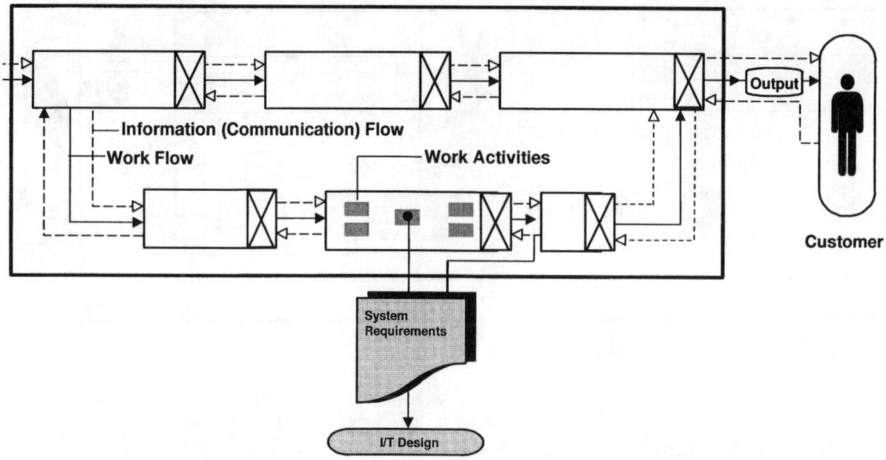

Figure 8.2. System Requirements Derive from the Business Process Flow

From the standpoint of the methods, tools, and techniques used, particularly when negotiating and reaching agreements and making commitments, they increasingly tend to work with information and knowledge; hence, they can also be identified as knowledge workers. The two terms are not identical but are certainly related. So, instead of being hired to do a specialized job, an employee is transformed into a process professional and is assigned a set of roles and responsibilities that directly derive from the work to be done in the process. Figure 8.2 shows how the roles and responsibilities of people working in the process are derived from the work flow *and* the flow of information within the process.

Through process reengineering, organizations are, in effect, reintegrating the work and information flow that Henry Ford and his contemporaries broke into elementary fragments almost a century ago. The vehicle of reintegration is the business process; its human executors are the process owner and the process professionals, as individuals, and the process management team, the synergistic group of process professionals who work together to achieve a common process goal shared by all. Process team members are not just representatives of the functional departments involved; rather, cross-functional process teams supersede the old functional departmental structure — just as the process professional replaces the employee specialist.

Even though work processes have been fragmented in a functionally organized corporation, the work still must be done. If we think of the progress of an invoice through an organization, we recognize that it gets handled by many different people — who carry out such tasks as authentication and validation, review and verification, and approval, among others. Organizationally, the people involved do not belong to the same department or function. They are scattered all over the organization — many times in different departments, groups, and divisions — the so-called organizational silos.

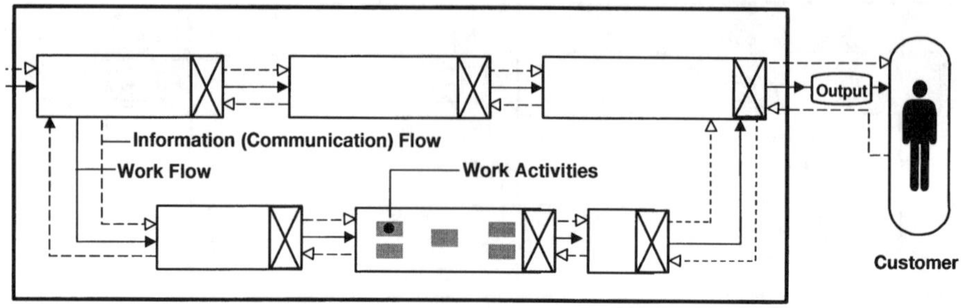

Figure 8.3. Integrated Business Process Flow

Process management looks at the same collection of people as a unit doing the work of handling, reconciling, and paying invoices — the invoicing process. Process reengineering does not necessarily change what they do; in fact, recognizing that they do invoice-related work, it aligns them so they can do it together in a more effective, synergistic way. It is in this sense that reengineering puts back together the work performed by a group of people who have been separated by functional organization and yet have to contribute to the same work product (e.g., the presentation of an invoice). They are rejoined by the appropriate assignment of a common goal and a shared commitment and thus become a process team. Many business process teams, organized to manage major product introductions, include key people from purchasing, marketing, development, sales, accounting, and other key functions — all located in a dedicated (physical or virtual) facility and working to make and meet commitments.

The process team is the work unit that naturally belongs together to perform a defined piece of work (the business process) and produce its output (the work product). The team approaches process work in an integrated fashion by virtue of simultaneously dealing with the information flow (communications) generated and the workflow performed in the process (Figure 8.3). For example, a shipping dock work unit needs to manage the information flow from the rest of the organization to help it meet the commitments to the customer. Failure to do so means nothing is shipped on time.

8.1.3 The Process Team

The skill most consistently identified as the critical skill for individual future success is an ability to become a member of a team quickly and successfully. The ability of a team to function successfully cannot be overemphasized. Experts predict that those who are unable to make the transition to a successful team member will find themselves unemployable. The interdependence of work expands daily because of the acceleration of change and its unpredictability. One human being, acting alone, will not be able to manage all the facets of a process.

Putting our astronauts into space is an awesome task. The complexity of the process is almost beyond human comprehension; it is certainly not routine, even though NASA has managed to make it seem so. Through the organization of the flow of communications and work around a series of commitments and go/no-go decisions (process centering), NASA is able to launch the astronauts into space on a consistent basis and accomplish their missions. Though the astronauts get the lion's share of attention, they are the first to admit that they are but one small part of a much larger process and a much larger team that have been engineered to work together.

Members of a process management team, who are collectively committed to process outcome rather than being individually responsible for specialized tasks, have a different kind of relationship to the overall process and to each other. They share joint responsibility with their fellow team members for doing the work of the process, not just a small piece of it. While each member of the team is doing different things — which, after all, is precisely why they have become members of the team—their mutual relationships largely eliminate the lines of specialization among them. Just through mutual contact and problem-solving, cross-training takes place. The shared commitment to process performance and its outcome and a basic familiarity with the sequence of major steps in the process make process team members generalists and their work multifunctional.

8.2 Process Work

The change to a cross-functional team as the new work unit results in a significant number of changes in the roles and relationships people have with regard to the process, the overall organization, its goals, and objectives. Turning our attention from work unit change to job change, we recognize that the change here is primarily in the set of responsibilities derived from the process work to be done by individuals. Jobs, which are a collection of specialized tasks, are transformed into process responsibilities that represent both multifunctional and multidimensional work.

8.2.1 Multifunctional Work

Functional work (whether on the assembly line, in the call center, or on the nursing station) is highly specialized and repetitious. It may require considerable training, such as how to connect the wiring of an ignition system or how to insert an intravenous feeding tube, for instance. The job may even require advanced education, such as a B.S. degree in civil engineering, in order to master the details needed to design the structural elements of a steel girder. But, when they perform specialized work, the employees usually do not know or even care about the overall process of, say, building a car, managing treatment for a potentially fatal illness, or designing a highway interchange.

Multifunctionality stems from the fact that process reengineering usually changes the boundaries between specialized kinds of work. At one credit company, for instance, credit checking for prospective borrowers was done in the credit department, pricing in the pricing department, and other terms and conditions in the business practices department — with the final credit approval provided by the manager of the customer credit department. When the customer credit process was redesigned, the four separate departments were abolished and all customer credit work was consolidated in the hands of multifunctional credit representatives. They were organized by type of customer served, not by the specialized work that needed to be performed.

8.2.2 Multidimensional Work

Process work becomes multidimensional when, in addition to performing activities with a functional content (administrative or technical, for example), people also assume the responsibility for managing commitments — for negotiating, decision-making, and communicating, among others. This work dimension requires good judgment and significant levels of knowledge, experience, and specific competencies.

8.2.3 Valuable Work

As work in the process-centered enterprise becomes more multifunctional and multidimensional, it also becomes more valuable and, as a result, more highly compensated. This occurs for two reasons:

1. Process reengineering eliminates not just plain waste but most of low-value-adding work, as well. Much of the coordination (counting, checking, reconciling, waiting, and tracking — the unthinking and unproductive work that became a necessity to compensate for process fragmentation) is eliminated by reengineering, which means that human capacity is freed up to spend more time producing more valuable products and services.
2. Through empowerment and the ability to make decisions and commitments, process professionals can leverage the productivity and responsiveness of others.

8.2.4 Productive Work

As a further result of process reengineering, work becomes more productive because employees perform a more substantive job (an activity or even a subprocess) that, by definition, produces a tangible work product that someone can measure and evaluate and that the customer really cares about. In this sense, process professionals become more like mini-entrepreneurs focused on the satisfaction of their immediate customers and not just following rules and procedures or abiding by a controlling bureaucracy.

8.2.5 Knowledge-Based Work

In a professional, knowledge-based environment, individual performance, promotions, and jobs are decided more and more by what people know and not by tenure, seniority, or position. While the notion of a knowledge-based company may have a bit of the abstract about it, there is nothing abstract about knowledge work. In knowledge work, information is the most important raw material one needs to do one's job. Knowledge work generates information and knowledge, and it increasingly leaves out routine, unthinking work, as well as the need for it.

A typical example of routine work is the old configuration of an airline company's accounting department with dozens of office workers sitting in rows at steel desks, sorting and counting used airline tickets so that they can match them up with travel agents' records to make sure the agents paid the airline what they owed. Another example is hundreds of clerks in the purchasing department matching purchase orders, invoices, and shipping manifests all day to determine which invoice was ready to be paid by the accounts payable department. Now all that mechanical, unthinking work has been automated; it is no longer done by people, but by computer systems that do it all automatically. Shipping docks scan shipping labels as goods are delivered and inspected, automatically instructing the accounts payable department to issue a check.

8.2.6 Rewarding Work

From a human resources and behavioral standpoint, work in a process-centered environment becomes more rewarding as people acquire a greater amount of learning, knowledge, and intellectual growth. In a business process team environment, individual recognition and development do not mean advancement up the hierarchical ladder; rather, they mean expanding one's breadth of experience and depth of learning so that one can assume the responsibility for increasingly larger parts of the business process. In a process-centered environment, jobs are not static, so continuous training and learning become integral parts of the job. Here, jobs become elastic assignments that can grow proportionately with the professional's experience and knowledge.

8.2.7 Work-Driven Shift in Personal Characteristics and Skills

Process-centered work is also more challenging and at times significantly more difficult. This is because process-centered work is much less procedural or routine; instead, it is knowledge work that requires creativity, the mastering of information to form judgments, and the ability to make decisions and commitments in the face of unexpected change.

Because of the transformed nature of process work, process-centered enterprises must consider additional criteria in their employment practices. In the traditional, task-oriented corporation, the prime criteria for prospective hires were education, training, and job skills. In a process-centered environment, on the other hand, where initiative, mental discipline, integrity, and empathy with the customer play important roles in making decisions and commitments, prospective employees' character and personality become important criteria, as well. By extension, this logic also applies to the internal training and professional development programs of the process-centered enterprise.

8.3 Process-Related Roles and Responsibilities

In the traditional, hierarchical company managed through command and control, most people played controlled roles. They were hard-wired into a hierarchical system of reporting and supervision. Their responsibility was to carry out specialized tasks, defined by formal job descriptions. In a process-centered enterprise where business process teams assume the responsibility of completing entire processes, team members' roles and responsibilities derive directly from understanding the implications of the integrated information and work flows in a given process. To succeed, they must be able to access the necessary information, make judgments and decisions, and translate them into executable commitments. To do this, they must be placed in empowered roles. Empowerment means more than permission or delegation; it means that members of a process team are required to interact, assess, use judgment, arrive at decisions, and ultimately make commitments which, in turn, will be fulfilled by those in the process.

It is logical to conclude that empowerment is an inevitable prerequisite of managing reengineered processes and of process centering on the whole. Every business process team must have the authority to make all process-related decisions — be it design-related or operational — within the boundaries of their formal obligation to the enterprise (agreed-upon goals, resources, schedules, standards, decision rules, work and time commitments, and the like).

Processes cannot be reengineered without empowering process design teams. Once reengineered, processes cannot be operated without empowering the process operations teams — without empowerment, they cannot make the necessary decisions and commitment changes to handle unexpected change.

8.4 Summary

The currently emerging process-centered structures operate in the context of conscious choice, which in turn is created through open communication and commitments freely negotiated and agreed to by empowered and accountable people — managers and process professionals. There are four specific sets of factors affecting the ability of the process professional to carry out his or her individual and team-related roles:

1. The management concepts and methods of empowerment, resulting in the ability to make commitments, which in turn lead to accountability — all key to the individual's ability to maintain the business process in a state of adaptability. We define accountability as the personal acceptance of the consequences of making a commitment. Only people can be held accountable; organizations, systems, processes, or machines cannot.
2. Work units changing from functionally differentiated departments managed by command and control to process teams operating as synergistic units, defined by the mutual and aggregate roles of their members. A special form of the process team is the process management team.
3. Transformation of specialized, functional work into multifunctional, multidimensional, knowledge-based process work. Work in the process-centered enterprise becomes more valuable and, as a result, more highly compensated. As a further result of process reengineering, work becomes more productive and rewarding as employees perform a more substantive job that the customer really cares about.
4. The individual process professional's roles and responsibilities derive directly from the process work he or she is expected to do. They replace the functionally specialized job of the hierarchically managed functional organization.

9 Process Centering: The Management Team

Because process centering is initiated, designed, and carried out by people, a look at the *who* is important before a more detailed discussion of the *what* and *how* of it can take place. Several management roles make up the enterprise management team that concerns itself with process centering the organization. An infrastructure of ownership is the basic mechanism for the assignment of accountability within each individual business process as well as throughout the process-centering effort.

9.1 Overseers and Implementers

Usually the cast of characters who attend to individual business processes (assessing, redesigning, and operating them) is different from the set of people who oversee the umbrella organization that ultimately concerns itself with the creation of the process-centered enterprise. Simply put, some people have unique roles that focus either on the individual process or on alignment at the enterprise level; others play dual roles as they carry out the tasks of bridging and integration at both levels. The environment in which they work (the process-centered enterprise) is illustrated in Figure 9.1. How the organization selects, assigns, and organizes these people is key to the success of process centering.

While there is no one-size-fits-all list of roles or designations, many years of reengineering experience have helped identify the following generic role set (Figure 9.2):

- *Enterprise transformation executive* — the senior executive agent of change who has the authority and resources to decide and motivate the overall process-centering transformation

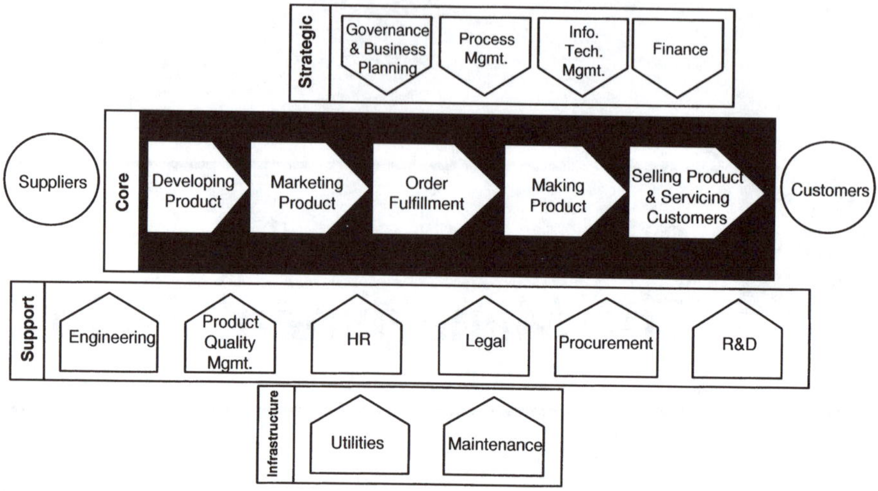

Figure 9.1. Process-Centered Enterprise Model

- *Enterprise transformation council* — influential policymakers who develop and champion the enterprise's overall process centering and business process reengineering strategy, commit the necessary resources, and oversee its progress; they are responsible for achieving synergy across the enterprise's separate, individual reengineering projects and for creating optimal leverage on overall business performance
- *Process management executive* — a senior manager who is the owner of the enterprise methodology of business process management and is responsible for developing and maintaining methods, tools, and techniques (including the latest information technology and computer software) related to business process reengineering and process centering
- *Business process owner* — an executive or middle manager with the responsibility and accountability for a specific, single business process, its reengineering, and subsequent operational implementation; there are usually a number of business process owners within an enterprise
- *Business process management team* — a group of selected individuals (usually managers or senior professionals) committed to the reengineering and subsequent operation of a specific, single business process; they define and analyze the current state of the process and design a new version or total replacement before planning its operational implementation
- *Business process management team leader* — a member of the business process management team selected by the process owner to lead and administer the day-to-day work of the process team
- *Business process stakeholders* — a body of senior managers with tangible stakes in the specific business process being reengineered and managed

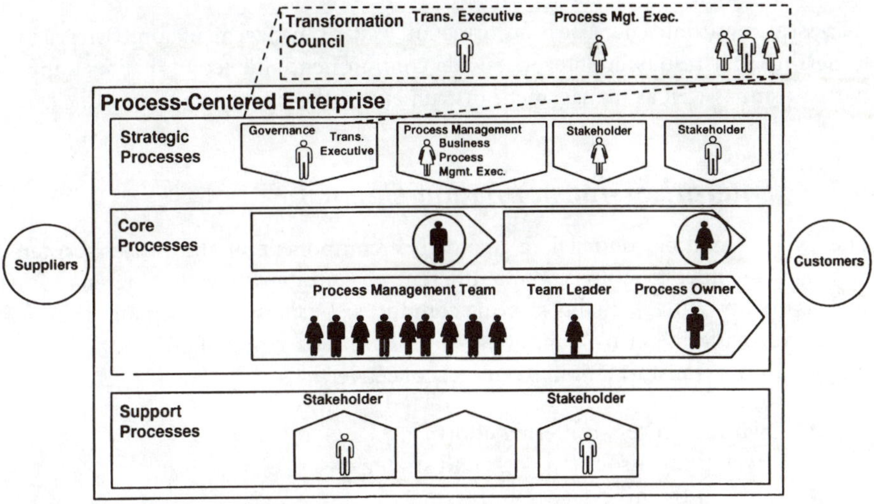

Figure 9.2. Key Roles in Process-Centered Management

As Figure 9.2 suggests, these roles do not represent a hierarchical set; rather, they are part of a two-level team structure where the transformation council and stakeholders form a strategic set of teams, and the business process management teams represent the operational set.

9.1.1 Enterprise Transformation Executive

The transformation executive is a senior executive with sufficient authority and power to make process centering happen. Because of the enormous impact of the change represented by process centering, this role is usually assumed, not assigned. In most cases of successful transformation, the transformation executive is the chief executive officer (CEO) or occasionally the chief operating officer (COO) of the corporation. Companies that start by naming a director, vice president, or even senior vice president of business transformation are already on the wrong track. Such staff executives, while having considerable persuasive and selling abilities, seldom possess the necessary clout to persuade the organization to accept the potentially radical disruptions and cultural changes that process centering may bring.

The transformation executive must be a passionate visionary and enthusiastic motivator with a clear sense of purpose who is capable of and emotionally and intellectually committed to staying the course and seeing the transformation effort through to its end. Hideo Sugiura of Honda, Bob Galvin of Motorola, David Kearns of Xerox, Roger Milliken of Milliken and Co., Jorma Ollila of Nokia, and Jack Welch of GE, among others, come to mind as being those transformation leaders who not only initiated and led major change efforts, but also stayed the course through to the successful, rewarding end. The transformation leader's management approach is active leadership that results in enterprise-wide commitment and action; his/her support mechanism, as well as management instrument, is the transformation council.

9.1.2 Enterprise Transformation Council

The transformation council is a mandatory component of the process-centered management infrastructure.[1] While different organizations give it different names such as quality council, quality steering committee, business transformation council, or transformation steering committee, it is essentially a group of policymakers who, along with the transformation executive, collectively:

- Own the process-centering effort
- Develop the associated goals and strategies
- Commit the necessary resources
- Monitor the ongoing progress of the effort to its completion

The transformation executive chairs the transformation council. The issues reviewed and decisions made by the council usually transcend the scope of individual business processes and reengineering projects; they normally have an overall, enterprise-level significance with a direct responsibility for overall business performance. The council acts as the final arbiter of inter-process issues and is the ultimate resolver of conflict among business process owners. Among the specific responsibilities of the transformation council are the following:

- Process-centering policy formulation
- Establishment of guidelines and rules for key business process identification and selection
- Identification and selection of individual business processes for reengineering; within this responsibility; the council (1) establishes the priorities among individual business process reengineering alternatives, (2) selects candidate processes on the basis of their relative economic value added, (3) decides what resources and time should be invested in these efforts and how, and (4) assigns business process owners
- Review and approval of individual business process mission statements, process goals, and performance commitments — many times formally summarized in process contracts
- Monitoring and assessing progress in all process-centering efforts

9.1.3 Business Process Management Executive

The transformation council attends to enterprise-level strategic aspects of process centering. Business process owners are focused on the redesign and operational performance of their individual business processes. The business process management executive, on the other hand, is the owner and manager of the formal processes and methodologies of business process management itself; in other words, he or she is the owner of the business process management technology.

The executive is an *ex officio* member of the transformation council who usually serves as the secretary to the transformation council and also as chief of staff to the business transformation leader. In terms of organizational structure, he or she should report directly to the transformation leader. In his/her active role, the business process management executive:

- Designs the enterprise's adaptable business process management process
- Maintains the cutting edge status of business process reengineering technology
- Establishes and maintains contact and communications with the process management community, external experts, professional associations, etc.

In his/her supportive role, the executive:

- Assists the transformation council in identifying and selecting individual business process owners and business process team members, as appropriate
- Oversees the use of reengineering technology with the view to maintaining integrity of method and consistency of implementation
- Provides assistance to individual process owners and process teams in their work, especially as it relates to training, the application of specific methods, and the use of tools and techniques

The business process management executive is also concerned with developing the infrastructure aspects of business process reengineering and, upon request from the transformation leader, acts as the coordinator of multiple, related reengineering efforts. In this sense, the executive may act as the project manager of the overall process-centering effort.

9.1.4 Business Process Owner

Business process ownership is a key managerial role and responsibility within the process-centered management infrastructure. Process owners are usually selected from the ranks with line responsibility who carry a certain clout within the organization by virtue of direct ownership and responsibility for a significant portion of the overall process resource. They have respect and credibility throughout the company. In the past, it has been the process owner's reputation and even career that were on the line; the process owner earned the credit for success and received the blame for failure. The new approach to the business process ownership is one in which there is shared responsibility. The process team succeeds or fails as a unit. Helping a team understand shared responsibility is a big task, as placing blame has been common.

What is key about the process owner is that he or she is accountable for the outcome of both process redesign and the operational performance of the reengineered process. The term "ownership" refers both to the outcome committed to and the authority over the means needed to achieve it. Accordingly, it is the process owner who will ensure the maximum adaptability of process design so that his or her individual business process, in concert with other reengineered processes, will contribute through their collective adaptability to the overall effective response of the enterprise.

When an organization first embarks on business process reengineering, good candidates for process ownership are few and far between. The best candidates usually come from management or team leader ranks, as both normally have many of the characteristics and experience necessary for the successful exercise of process ownership. Some process-oriented companies have used their previous total quality management experience and specifically developed and groomed individual project team leaders to become future business process owners. Conversely, other companies have used process ownership as the grooming mechanism for future senior-level positions, general managers, and executives.

The business process owner is both a leader and enabler. Especially for larger business processes, the owner's job is to select the business process management team and enable it to do its job — be it the redesign of the process or its operational management. Rarely do owners do the daily, hands-on work of process reengineering and operations; it is the business process team leader who receives that assignment from the process owner. The five most important enabling tasks of the business process owner are

1. *Task One.* Manage the customer-supplier relationship as it is related to his or her business process (the timely and complete negotiation and communication of commitments), both within the process and to the rest of the organization.
2. *Task Two.* Obtain the necessary resources for the business process to satisfactorily meet commitments.
3. *Task Three.* Motivate and advise (in the professional and managerial sense) the process management team.
4. *Task Four.* Build and maintain a productive relationship and enlist the support of the transformation council and, even more importantly, the business process stakeholders.
5. *Task Five.* Gain the cooperation of other business process owners and attend, as much as possible, to issues and the resolution of conflicts.

Unless reassigned to another process, the process owner's job lasts through the lifetime of the business process; consequently, it will not end when, for example, the reengineering phase of the overall management of the business process is completed. Just like the business process management team, the process owner will continue to attend to and be accountable for the operational performance of his or her business process.

9.1.5 Business Process Management Team

The hands-on work of reengineering and the day-to-day operational management of the process are the responsibility of the members of the business process management team, individually and collectively. These are the managers and professionals — and no one else — who must produce the ideas, judgments, decisions, and plans for redesigning or completely replacing an existing business process. And, when all of that is done, they are also asked to implement it, to make it an operational reality. These are the only people who can, in the final analysis, reinvent the business and cause it to respond to all possible change through the cumulative adaptability of the business processes.

Several rules have emerged for the selection and deployment of a business process management team though years of reengineering experience and from lessons learned in success and disappointment:

1. *Rule One.* Every business process management team is dedicated to a single process; no team can reengineer or manage more than one process at a time.
2. *Rule Two.* A business process management team is a synergistic body of people in the true sense of the word. Process teams are not groups, committees, or task forces. A smoothly functioning, productive, and effective core team which stays in place through the lifetime of the process is usually small (between six and twelve people, even though its incumbent members may change) because stability of membership is crucial to long-term success.
3. *Rule Three.* During certain phases of the reengineering effort, the core team is expanded to become the extended process team which includes people from outside the process (namely, a few of its customers), as well as people from outside the enterprise. This extended team, therefore, grows and shrinks with the expansion and contraction of the work placed upon it.
4. *Rule Four.* Membership on a business process management team is neither a part-time nor a short assignment. Depending on the current functional responsibilities of team members within or outside the process, it is recommended that all team members make a minimum time commitment of 75% of the members' time to be spent on the process team or with work directly related to its assignment. Outsiders to the process will be able to get away with a lesser obligation, close to 50% as the upper limit. Insiders are expected to make a full commitment, if at all possible. This is based on the practical fact that most members of the core team are already spending most of their time with process-related work. Full-time participation makes accomplishment of the reengineering or management project easier and faster; a high-level commitment also sends a powerful signal to the rest of the organization that its management is serious about, and fully committed to, process reengineering and process-centering.

Rule Three warrants further discussion. It is clear from the foregoing that every business process management team has two kinds of members on it. Hammer calls them *insiders* and *outsiders* — as related to the business process.[2] Insiders are managers or senior professionals who currently work in one of the functions of the process that is undergoing reengineering. Their prime contributions to the business process management team are

- Process knowledge
- Credibility and familiarity with co-workers in the process
- Understanding of problems and what needs to be changed

One limitation of insiders is that they are many times tempted to look ahead at what the process should be, as opposed to first looking at what it is now. This is because the individual perspectives of insiders may be limited to just one part, one subprocess within the entire business process. They may also hold a vested interest in the *status quo* and their own position within it. Insiders display a higher propensity of staying within the framework of the existing process, rather than discarding it. Many companies which have assembled business process teams made up entirely of insiders have come to realize, with regret, that such teams have essentially recreated a modified version of what had already existed — with perhaps an incremental improvement.

Outsiders, on the other hand, do not currently work in any of the functions that are part of the process, although they may have done so earlier. Outsider candidates for business process management should be people who have a long enough experience with the enterprise to know the ropes in the organization and who bring a higher level of objectivity and an independent perspective to the process team. If anyone, it tends to be the outsider who thinks "out of the box" and who brings new and unorthodox ideas and even asks naive questions to challenge existing assumptions and to make waves. Outsiders include specialists with experience or skills in particular disciplines such as law, human resources, information technology, accounting, and the like. They have the special knowledge and information that the team needs at certain points in the reengineering effort. As a consequence, they are usually assigned as on-call, *ad hoc* or part-time team members.

The first choice for outside members of the business process team is customers of the process. Other good places to look are overarching, common functions such as information systems, finance, and marketing. Qualified consultants can be engaged to amplify the beneficial role and influence of outsiders, especially when a company does not have access to appropriate candidates within its own boundaries. These consultants can be primarily involved in training and facilitation roles to support the business process management team and, on specific occasions, work with the transformation council, as well.

There is no hard and fast rule for the ratio of insiders to outsiders, as this depends on the process under consideration, the choice and availability of candidates, and many other factors. On a business process core team with a total membership of ten, two to three outsiders would probably provide the right ratio.

The business process team's overall performance and its progress toward its goal are measured and evaluated by the process owner — not so much as a boss but more as the team's customer. The performance of individual process team members is measured in the context of overall team performance.

As we already have said, business process owners, supported by their teams, are fully accountable for the outcome of their individual business process. For this reason, both the process owner and the process team, as entities, stay in place through the lifetime of the process, which, in practical terms, may mean many years. And, while the incumbents of process ownership and members of process teams may change over time, process team members should remain on the team at least through implementation of the planning done for the business process they have reengineered, if that is where their assignment started. For the average size business process in business and industry today, this usually takes 9 to 15 months.

Our recommendation, borne out by many years of practical experience, is that both the process owner and his or her core team should take the reengineered process well into its operational phase and gain the invaluable experience of operating a newly redesigned or brand new business process.

9.1.6 Business Process Management Team Leader

As we have said before, business process management teams are empowered and, therefore, by definition, self-directed. As such, they do not have an official manager or boss — although it is an almost universal practice that business process reengineering teams will have a team leader appointed by the process owner. In some companies, business process teams are allowed to nominate their team leader by acclamation. This, however, is a practice we do not generally encourage because popularity does not always mean having the skills and competencies to perform the team leader's challenging role. The team leader, as we said earlier, is not the boss but first among equals to provide the necessary organization, coordination, and administration for the team's continuing work. (The team leader's responsibilities are described in more detail in Chapter 12.)

9.1.7 Business Process Stakeholders

The heads of functions such as sales, development, finance, information systems, and so on represented on the business process management team are the key stakeholders of that business process. For large complex processes, obtaining buy-in and ongoing support from these functional executives is crucial if the process management team's efforts are to be successful. If necessary, help can be obtained from the transformation council to gain all of the necessary cooperation from these stakeholder executives, especially as some of them may also be among the council's members.

The process stakeholders are, in essence, allies of the process owner and his or her team. Process stakeholders should support the owner and the process management team in all phases of the reengineering effort. This is especially important when the process scope must be changed to include new functional capabilities necessary to satisfy new requirements, or when a particular tactical improvement project requires the knowledge and expertise of a functional manager not on the original core team.

9.2 Process Management Resources

The majority of human resources required for the reengineering effort is provided by the various functional areas of the organization. The functional heads of these areas are usually the stakeholders of the process. The means of securing the resources is some mechanism of management agreement. The needed skills are then acquired primarily by means of training and, later on, through ongoing on-the-job work experience.

9.2.1 Process Contract

One frequently used mechanism for securing and documenting the process stakeholders' support is the *process contract.* It is a formal agreement between the process owner and the process stakeholders, usually established at the time the process owner is appointed by senior management. The process contract — sometimes also called the process charter or process operations memorandum — secures the essential resources needed by the process owner to carry out the process reengineering effort. It is the management mechanism that establishes the process stakeholders' specific obligations to support the management of the business process in terms of (1) resources, and (2) participation in process-related reviews and decision-making. The process contract documents include:

- The process mission, initial goals, and assumptions about the desired outcome
- Specific understanding as to the scope and organizational boundaries of the process and, in particular, the customer-supplier relationships it comprises
- The authority of the process owner and the process team members and the rules for resolving inter-functional issues
- Specific contributions, primarily in terms of human resources, by the various functional areas
- The general methodology and schedule to be followed by the process owner and the process team, with particular emphasis on stakeholder involvement, reviews, updates, and decision points

To fulfill their obligations, stakeholders need a basic orientation on business process management and, in particular, on the methods of reengineering. Beyond this, we cannot overemphasize the importance and value of good business rapport between the process owner and the stakeholders of the process. One example of a well-working owner/stakeholder relationship (and its beneficial effect on the successful completion of the related reengineering effort) is Pfizer Central Research Division's reengineering of its procurement process in the late 1990s. From the beginning, the process owner set up a disciplined and consistent roadmap of stakeholder participation in key reviews and decisions throughout the project. This paid off handsomely when funds had to be approved for the new information systems necessary to support the advanced procurement environment called for by the reengineering.

Another success story is Union Camp Fine Paper Division's order-fulfillment process. Here, again, the process owner and the process team leader included critical milestone reviews and decision points in the project plan to specifically ensure the active participation of process stakeholders, who, in essence, represented the entire divisional executive management. For example, stakeholders' involvement in developing the project communication plan resulted in the informed acceptance of the reengineering goals by virtually all the division's employees. Similarly, a series of carefully planned stakeholder briefings and reviews led to the quick approval of considerable capital investments needed for the newly reengineered process and its enterprise resource planning system implementation — provided by the Dutch software company Baan.

9.2.2 Process Training

As detailed in the previous chapter, process-centered enterprises must consider additional criteria in their employment practices. This logic also applies to the internal training and professional development programs for the process-centered enterprise's management team. The most important human resources of a business process are the process owner, the business process management team, and the process stakeholders. This group of people constitutes the human resources infrastructure of an individual business process, even though other members of the enterprise process management or various functions may also get involved from time to time. Accordingly, appropriate initial training and ongoing individual and team-related development of the process owner and process team members are of paramount importance to the effectiveness of business process management.

The core team, which is made up almost exclusively of insiders, essentially stays in place for the lifetime of the business process. The training requirements of the incumbents will depend on the length of time they are committed to stay on the team and on the particular phases of process management they participate in; therefore, the training requirements of core team members will always include the use of collaborative tools and techniques. Beyond this, however, they will need training in process design methodology and supporting tools (e.g., process description and mapping, process analysis, data collection and interpretation techniques, activity-based costing, and project management, among others). If the team is intended to stay together for a longer period of time, its members will also benefit by training in team building, meeting management, and interpersonal communications.

The extended team, which includes a larger number of outsiders, stays in place for much shorter periods of time in the same composition. Therefore, their training requirements as a team are focused mainly on the use of collaborative tools and techniques. For example, all members of Union Camp Fine Paper Division's extended order-fulfillment process team were given comprehensive training in the messaging and database sharing systems used by the company, such as Lotus Notes.

The business process management team leader essentially runs the process team's work. To be effective, the team leader's formal training should include team facilitation and leadership skills, in addition to receiving all the training that team members also participate in.

9.3 Summary

The successful management of individual processes and process centering is predicated on leadership by accountable individuals, leading well-trained, competent, and experienced process teams. The functional organization is partially replaced by: (1) a set of cross-functional business process management teams operating at the individual process level, led by their respective process owners, and (2) an organization-wide process management infrastructure for the coordination of the various reengineering efforts and for overseeing the process-centered transformation of the enterprise.

Process-centered management represents a unique infrastructure, including a set of key managerial roles, some enterprise-wide (transformation leader, members of the transformation council, process management executive) and some related to individual business processes (process owner, members of the process management team, team leader, process stakeholders). Within the management structure, the importance and value of good business rapport between the process owner and stakeholders cannot be overemphasized. The hands-on work of reengineering and the day-to-day operational management of the process are the responsibility of the business process management team.

The human resources necessary to support this infrastructure from within the organization are secured primarily through commitments from the process stakeholders. One frequently used mechanism for documenting the process stakeholders' support is the process contract. The most important human resources of a business process are the process owner, the business process management team, and the process stakeholders. Appropriate initial training as well as ongoing individual and team-related development of these people are of great importance to the effectiveness of business process management. External resources are acquired and developed through an appropriately enhanced employment and training process.

10 Information Technology: The Response Integrator

Good, timely information and the people using it make processes adaptable. Though the human mind remains superior intellectually, information technology allows each individual to accelerate his or her capacity to assimilate and manipulate data when dealing with the unexpected or unpredictable. This gives the data meaning and value for sound decision-making in the face of change. It is technology that allows the fast evaluation of available options before a commitment is made to handle change. And it is technology that makes possible the quick deployment of needed capabilities and the rapid execution of process activities called for. This ability of technology to help integrate commitment management and related workflow is the new dimension that extends far beyond the traditional scope of enabling.

10.1 Change and Information Intensity

As we mentioned earlier, adaptability of any human organization, enterprise, or process is primarily dependent on its information intensity — the breadth and depth of information content and flow. Achieving adaptability, then, is primarily an information management issue. In the face of unpredictable change, some of the most valuable information is that generated in exchange with customers. All this means is that adaptable business processes in a process-centered enterprise — and the enterprise itself — must be made information intensive:[1]

> *A process is said to be information intensive if its execution is based on the information received, processed, and communicated along the chain of customer-supplier relationships within and outside the process. An information-intensive process can, by definition, change as it receives and absorbs new information from its environment.*

10.1.1 Information Technology

An important purpose of information technology in a process-centered environment is to provide information that helps management determine the appropriate response to unpredictable change. The tangible results of decision-making are *commitments,* which will, in turn, activate the various subprocesses and activities of a given business process, but this cannot happen if the information is incomplete, incorrectly evaluated, or never communicated among key people. Accordingly, one basic information activity in an information-intensive environment is the ongoing translation of all incoming data into usable information for the purpose of distribution and sharing in a specific process. Keeping track of the status of all the commitments within the process is the prerequisite of effective commitment management. It requires a highly responsive management tool that only information technology can provide.

10.1.1.1 Capability Integration

Another key purpose of information technology is to help management execute a chosen response through the rapid selection, deployment, and activation of appropriate process capabilities. This managerial action is known as *capability integration.* From an information management viewpoint, capability integration is the definition and assembly of relevant information about functionally and geographically dispersed subsystems (procedures, activities, and related resources) for a given purpose. This organized information frequently takes the form of a process flow model — the management tool used to guide the execution of process activities through completion (meeting the conditions of satisfaction). Various technology-based tools such as messaging, data warehousing, and process mapping systems, as well as workflow managers, help process owners and their teams do their job.

Translated into everyday terms, capability integration means that if a customer has a defective power tool that cannot be repaired at the manufacturer's local repair shop, the best solution would be to send it to the closest service center which may be 200 miles away. The service center is set up to turn around 95% of the repairs in a week or less. If it is determined that it will be longer, thus causing major disruption for the customer, or if there is a strong warranty issue, the center manager is empowered to loan the customer a tool, if possible, or replace it immediately. The customer is satisfied, and a prolonged process is avoided that would consume more resources, manpower, and money than the cost to replace the tool. The information technologies that track commitments and integrate capabilities derive their initial specifications from the information content of the business process (Figure 10.1).

10.1.1.2 Informational Representation

As mentioned earlier, process reengineering helps eliminate low value-adding, routine activities and the resources expended carrying them out. In traditional, functionally specialized organizations, it has always been necessary to coordinate the fragmented

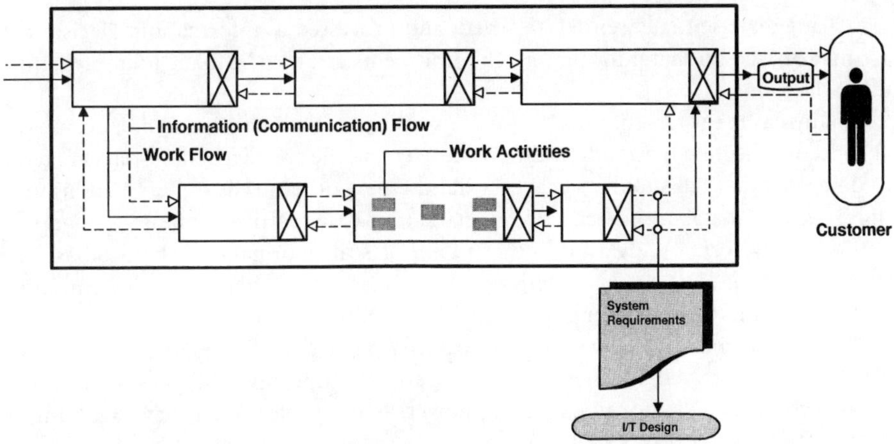

Figure 10.1. System Requirements Derived from the Business Process Flow

information in a functionally specialized organization. In the emerging process-centered environment, the role of coordinating activities (such as the various overhead functions of expediting, inspecting, controlling, etc.) and the related human resources are simply replaced by high-quality information.

Recent advances in information technology have shown that it is possible to create robust informational representations of change that are occurring inside or outside the organization. Systems software is aiding managers in this effort. Interpreting and analyzing the change occurs by evaluating this information. It is also possible to manage complex business processes (and, by extension, process-centered enterprises) by just managing their informational representations. As a result, a process can be manipulated (changed) by manipulating the information that describes it.

Individuals who are using one of the many online brokerage services, such as Charles Schwab or Fidelity, are able to manipulate their stock portfolios based on minute-by-minute stock market information. With new systems installed on Wall Street and linked to expanded capabilities within the brokerage houses, the process of real-time buying and selling appears almost seamless to the customer.

10.2 Information Management for Adaptability

Information management in a process-centered environment means that management can organize resources into tangible structures (capabilities) and executable actions by simply manipulating the information representing them. A car buyer can now sit down in front of a personal computer in his or her home and assemble an electronic representation of his or her car of choice, trying out all the features, options, and colors available. Once the desired configuration is attained, an order can be transmitted to the car dealership for delivery.

Change is essentially sensed, received, and processed as information. Decisions are made based on that information. Commitments can then be made to implement those decisions.

Many cash registers in grocery stores are set up so that when an item is scanned at checkout the item is automatically reordered from the supplier. All the data are fed to databases that tell suppliers exactly what the next order needs to be. In addition, those reorder databases often supply marketing data, as the computer program automatically sorts the data into other usable sets of information. The database is even capable of alerting the supplier of an increase in popularity of an item and seasonal buying trends, among others.

When information changes (as when external change accelerates), new structures may have to be created. Retailers, for a long period of time, missed the fact that those 16 and younger had enormous buying power. Now retailers often have age range data-entry keys to capture that important data at their cashier counters.

10.2.1 Two Key Process Components

In an operating, but adaptable and therefore potentially changing process, *information* is one fundamental component. For a process to remain productive and evolve into higher levels of organization, information must be continually generated in usable forms. Information is unique as a resource because of its capacity to multiply or expand itself, depending on how it is used.[2]

As mentioned in the opening of the chapter, information cannot intellectually transform itself without human intervention: a set of data needs interpretation to transform itself into information that now has another use. Such interpretations can only be made by humans; therefore, the other key component of the adaptable process is the *human* one.[3]

By managing its information content such as inputs, process activities, performance measurement, and control data, a process can be made into a well-ordered open system, capable of generating, absorbing, and transforming information.

10.2.2 Basic Functional Capabilities

Information management, as described here, allows for maximum process adaptability and enterprise responsiveness to change. At a minimum, the underlying information management architecture must include three basic functional capabilities, all able to deal with various symbolic representations and models:

- Information sensing (capture, or the ability to access and store all that is available and relevant)
- Information transmission (communications, or the acts and means by which it is sent and received)
- Information processing (manipulation and transformation, or the evaluation and determination of its uses)

In terms of a simple example, an organization would be able to systematically capture the information that employees working in the field often are aware of. A field service representative might hear about a large project that is about to be let for bid. Though it is not in his or her area of responsibility, it is still something that the representative's company does. The company would have a system in place whereby that information is fed into a potential sales lead database that communicates it to the sales force within a few minutes. The result is that the salespeople are alerted, and they contact the prospect, ask about the bidding process, and are successful in getting a new piece of business simply because one of the employees knew how to transmit the information through the company's information system. Sometimes called cross-selling, this approach is becoming institutionalized in many enterprises with the help of technology.

In the process-centered enterprise, these functional capabilities must be shareable within and among business processes, across the enterprise, and even with other enterprises (suppliers, business partners, co-manufacturers, etc.). Accordingly, one of the most important and valuable enabling and integrating technologies in the process-centered environment is that of network-connected collaborative systems.

10.2.3 Technology Assessment

Understanding what is presently available in information technology is a start. Competitive organizations will have assigned people spend all their time scouting new technology and making recommendations about what uses it has in the organization's future. Technology assessment may become one of the hot new jobs of the 21st century. People who do this well are worth every penny of the high salaries they are commanding. In addition, they have the uncanny ability of accurately figuring out how the technology will enhance their company's processes and strengthen its ability to adapt and keep all parts of the enterprise responsive. Unfortunately, many companies have not understood how critical ongoing technology assessment will be to their future success.

Another key technology in the process-centered enterprise, described in more detail in the last chapter of this book, is known as the enterprise resource planning (ERP) system. The name derives from the purpose of these systems: to plan, organize, align, and operate the enterprise's resources in an efficient way within one or across multiple business processes. Having a professional on a process team who does nothing but assess technology and design its applications for the process is the only way a business process can become fully responsive to rapid and unpredictable change. Many process management leaders and reengineering practitioners are using technology assessment professionals as an integral part of their process teams.

10.3 Networked Collaborative Systems

One of the most exciting approaches to come along is the innovative use of an organization's internal information network (or networks connecting several organizations) to deliver highly customized functionality (online training, job-related data, and work products) directly and simultaneously to each participating employee's desktop or laptop computer. Networks connect people and make possible information exchange and work sharing, as well as collaboration among those connected. Over the past decade, some major developments have contributed to the growth of such participative, online collaborative systems:

- The significant increase in the internal training needs (technical, business, and professional) of organizations in both the public and private sectors and the availability of authoring software that allows even novices to design and set up interactive training and relatively simple, cost-efficient, online training programs
- The advent of process-oriented work practices based on collaborative group structures such as project teams, empowered cross-functional teams, and self-directed work teams
- The development and installation of wide-area networks (WANs)
- The emergence of groupware, the class of systems and software that enable efficient communications and the shared implementation of work activities within and among work teams

10.3.1 Employee Training

Over the past two decades, and more than ever before, internal employee training has become a key strategy and, in many cases, a prerequisite for achieving business success. Many organizations in the U.S., Japan, and Europe have launched massive training initiatives in the basics of business and technology, quality, and other strategic disciplines. One of the Fortune 50 companies is installing a project management system to be used by all project teams in the organization. Preparatory to its implementation, all those involved have been given an interactive training compact disc on project management basics that has a pre-test, tutorial, and post-test. The results must be sent back to the training function, which assesses the level of knowledge of each person based on the results of the pre- and post-tests. In addition, the compact disc records how long it was in use.

According to one report,[4] the education and training costs of U.S. businesses are now estimated at over $30 billion per year. The Saturn Division of General Motors provides its 9000 workers an average of 90 hours of training per year. Saturn expects all employees to spend about 5% of their time in training and education each year. Intel University now has a training budget equal to about 5.7% of company payroll. Motorola University has a budget of $120 million per year and 14 branches in the U.S., with more branches in Europe, Asia, and Latin America. IBM has a standing policy requiring each and every employee to receive a minimum of 40 hours of training every year.

Many companies have developed their own in-house, world-class education programs with related professional training organizations and support systems for their employees. Some claim that they now provide over 80% of the training for their employees in-house, whereas a few years ago, much more training was done by schools, consultants, and other outside providers.

10.3.2 Collaborative Work Practices

The application of collaborative technology to a variety of work shared by a number of individual or group participants results in collaborative work practices such as the joint design of a process layout, team-based implementation of a project, and the like. Collaborative work practices are based on the team as the basic work unit. The team may perform work in two modalities: (1) learning as a team, and (2) processing as a team by executing tasks or activities learned, resulting in a specific work product. In either mode, participants are organized and geographically located either as individuals (located in many places) or teams. A larger team may be made up of several smaller sub-teams or squads, each located at a different place.

In a process-centered environment, these practices represent an increasingly useful, effective, and highly time- and cost-efficient way of carrying out shared process work by participants who are geographically dispersed or when schedules conflict. They allow specially trained experts (trainers, facilitators, team leaders, or process owners) to teach and work on advanced solutions together with other professional participants at their own desks in multiple locations.

Both individual and team participants may work in one or both of two timing relationships: synchronous (all simultaneously, such as via teleconferencing or chat rooms) or asynchronous (at different times, usually of their own choosing, such as by means of e-mail or an electronic bulletin board designed for this purpose). Each of these relationships offers specific advantages to the participants and to the organization implementing collaborative work practices. In fact, collaborative solutions are implemented using a mix of these geographic and timing strategies (see Figure 10.2).

10.3.3 Wide-Area Networks

Wide area networks are dedicated, high-speed network links that interconnect a series of local area networks (LANs) within an organization or even among several organizations. The purpose of WANs, which normally use proprietary communications protocols to run a large variety of standard or customized applications, is to make information more widely available across geographic and organizational boundaries and thereby facilitate collaboration among several groups, sites, or processes.

The virtual office uses WAN technology. Companies have found that when they send their sales people out to the customer, they can continue to communicate with the company's databases from any place that has a modem hook-up — which now includes cellular telephones. The new digital cellular phones offer sophisticated security, making communications not only easier but also more secure from an intellectual property and privacy standpoint. Sales people are no longer isolated. They

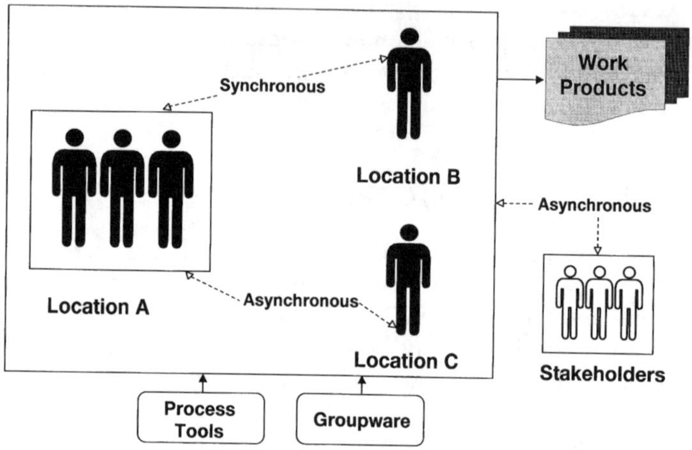

Figure 10.2. Collaborative Work Practices

download data and use it at the point of sale; no more rash promises about product availability or delivery. In addition, important followup documents can be generated, letting customers know that they are important and their accounts are being given priority. The virtual office is one in which a professional can function any time, any place, and anywhere because information technology allows it to exist.

10.3.4 Groupware and Collaborative Computing

Collaborative computing is the information technology network and software applications that support work teams. While these systems enable asynchronous and geographically dispersed efforts, collaborative computing supports more traditional teams as well. The system trains team members on a given subject (e.g., process improvement, programming, and project management, among others), provides tools to support the subsequent team efforts, and allows archiving and retrieval for current and future projects whenever a team member needs it, wherever that team member is located.

10.3.5 Fundamentals of Online Collaborative Systems

The organizational computing model has migrated from centralized mainframe computing to decentralized online workstations and now to a scenario with powerful local workstations connected to each other and to common databases. The phenomenal power at the desktop or laptop, combined with increasingly fast connections and creative software solutions, is just now allowing truly collaborative computing to emerge (see Figure 10.3).

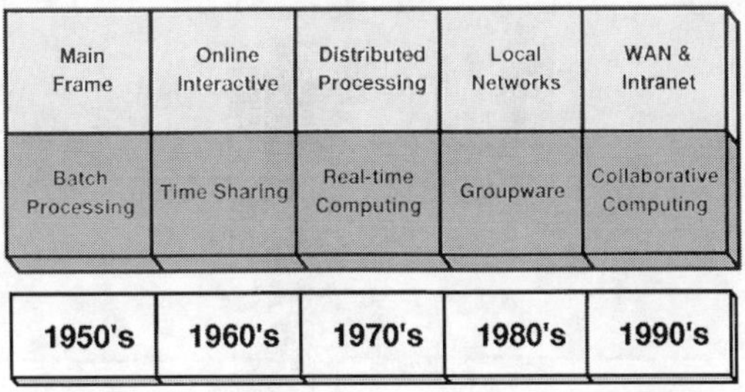

Figure 10.3. Evolution of Computing Technologies and Applications

Every organization has specialized needs to carry out collaborative work. What are some examples of collaborative computing?

- Design teams strive to develop new products across time zones to drastically reduce cycle times.
- Geographically dispersed sales organizations need a common strategy and facilitated access to marketing, account profile, and pricing information.
- Process design and implementation teams develop reengineered processes for worldwide deployment and operation.
- Manufacturers and service organizations develop new relationships with their vendors and customers.

So, each organization's specific collaborative computing criteria differ, but they are likely to include some common elements. A common collaborative database to capture and archive information is the foundation of any collaborative system. An intuitive, easy-to-use, and easy-to-access user interface is another critical success factor. Videoconferencing and online whiteboard capability are approaching sufficient maturity to be included in collaborative computing designs.

In summary, collaborative computing is an advanced version of distance learning and distance work sharing. At the same time, the applications supporting the system can be as diverse as the roles of the users:

- Intellectual capital systems for knowledge organizations such as legal, marketing, or sales
- Process mapping for process reengineering teams
- CAD for design engineers
- Imaging applications for advertising and communications

10.3.6 Collaborative Systems Architecture

The architecture of a well-designed collaborative system includes a front, middle, and back end (see Figure 10.4).

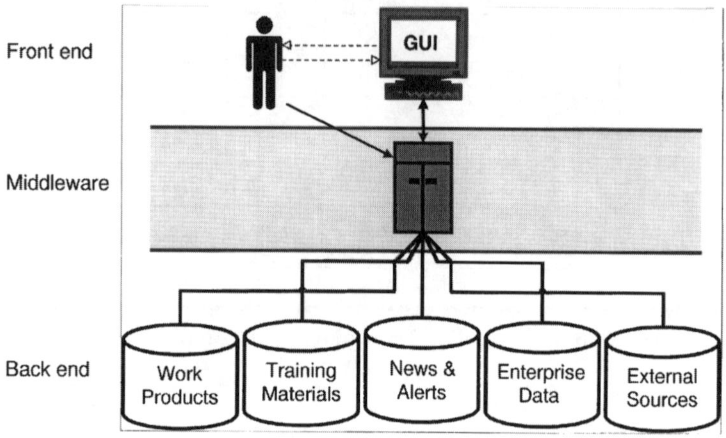

Figure 10.4. Collaborative System Architecture

10.3.6.1 Front End

The hardware and software at the front end, or graphic user interface (GUI), provide the visual, audio, and tactile relationship with the user. Techniques such as animation, voice-over, video, and effective design can further enhance user communication. The interface design should add measurable value to the individual user by following these rules:

1. *Rule One.* Keep it highly interactive.
2. *Rule Two.* Avoid putting textbooks online (common mistake).
3. *Rule Three.* Utilize active information distribution approaches known as push technologies.
4. *Rule Four.* Personalize to meet individual needs and preferences.
5. *Rule Five.* Include multimedia (sound, animation, and video).
6. *Rule Six.* Automate user support (developing multimedia "how-to" that reduces the need for a help-desk function).
7. *Rule Seven.* Integrate with groupware (e.g., Lotus Notes, MeetingKit, Team Quest) and relevant databases.

10.3.6.2 Middleware

Middleware provides the intelligence of the system. Profiling or individual passwords and log-ons allow the system to anticipate user needs and present information on a current project or promote a new online training module. The middleware also tracks individual work practices, such as navigation, and provides usage reports as appropriate. Information systems support people are able to anticipate overload problems, under-utilization, and other related issues.

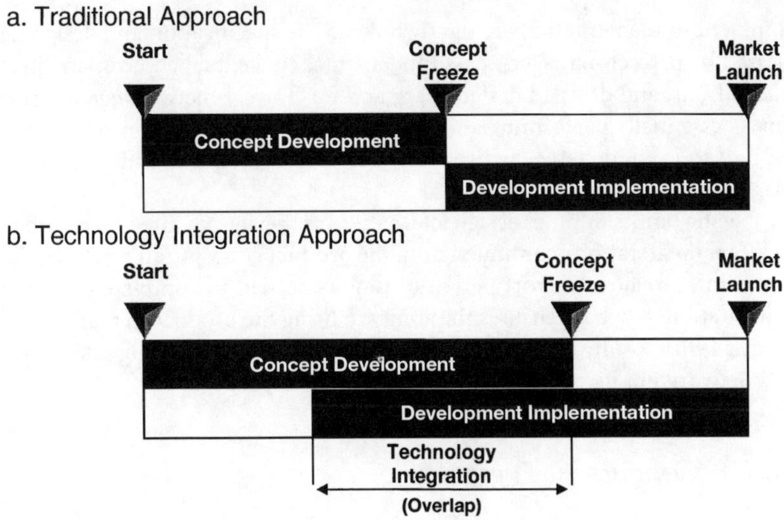

Figure 10.5. The Product Development Process

10.3.6.3 Back End

The back end usually consists of new databases that are created to support the collaborative computing system or existing databases and feeds, either within the organization or on the Internet.

These three architectural components—front end, middleware, and back-end—work together to support the success criteria identified in the initial system requirements. A good example of capability integration, supported by collaborative systems, is the new and flexible approach to product development using technology integration.

Traditional product development was built around a highly structured process that comprised clearly defined sequential phases. IBM's phase process of the 1970s for the development of new hardware and software products was a typical example of this approach. Its basic philosophy was to minimize uncertainty and to avoid unnecessary change during development. To this end, the development process was essentially divided into two segments: (1) concept development, which yielded a stable product concept, and (2) efficient, consistent development implementation, resulting in a product ready for market launch (Figure 10.5a).

This approach worked well as long as technology, customer needs, and competitive developments were reasonably predictable. In today's fast-moving, technology-driven industries, however, product development must embrace rather than exclude change. This is done essentially by allowing the concept development segment to overlap with development implementation as long as possible and by integrating new knowledge and the various emerging technological changes into the design on an ongoing basis (Figure 10.5b). During the overlap phase, concept development and development implementation are tightly linked. The specification and design teams, supervised by project managers, pass fundamental conceptual choices, architectural representations, and detailed designs back and forth over a networked collaborative system — essentially performing an iterative technology integration. This is needed to respond to unpredictable changes in technology or market conditions while the development effort is still underway.

At specific points in time, design alternatives are reviewed, integrated, and tested — and then the iterations continued until the product concept can finally be frozen. At this point, a relatively short period of time is needed to complete development implementation — which ensures the ability to bring the product to market without much delay. Microsoft, Netscape, and other information technology developers of today's network era use this product development approach.

10.3.6.4 Concurrent Engineering

The overlapped implementation of the development process is also known as *concurrent engineering,* where the prime purpose is not just to save time but also to integrate the technical and commercial attributes required for a product to be competitive during its *entire useful life.* This is achieved through a systematic approach of meeting customers' requirements by reaching the earliest possible agreement on those requirements and subsequently completing design, manufacturing, quality assurance, cost and schedule management tasks *in parallel.* In other words, key to the success of concurrent engineering is making and then meeting the early commitment to the customer through the disciplined, parallel management of the design, production, and post-production processes.

The "up-front" integrated approach has been shown to be critical, as most of a product's quality problems and associated costs and time delays arise directly from the design process. U.S. Department of Defense studies show that concurrent engineering and related work practices can reduce design changes by 50%, development time by 40 to 60%, and defects, scrap, and rework by up to 75%.

Concurrent engineering is primarily used in the automotive, aerospace, high-technology, and software industries. It depends on close teamwork among all company functions involved in a product's lifecycle and engages customers and suppliers in the entire product introduction process. This offers an ideal environment for the combined implementation of both collaborative systems and ERP solutions.

10.4 Information Technology as Integrator

In the emerging information-intensive world, new information technologies and applications are developed and made available at an ever-increasing rate. In the face of continuous and unpredictable change, the management of the process-centered enterprise must always be prepared to recognize and then take advantage of the integrating potential of new or changing technologies. To do so, they must develop a systematic way of reasoning in their decision-making that will lead to a reasonably consistent set of choices of enabling technologies.

10.4.1 Deductive Thinking

Deduction is reasoning from the general to the particular. From already known principles, new hypothetical principles are formulated. These principles are tested by particular applications; as the results of the tests satisfy the conditions of the hypotheses, laws or rules are arrived at, and from these laws future results may be determined. The process of deduction is considered infallible if the original proposition (that is, the known principle) is true. Deductive thinking is the approach of the problem-solver; here the risks are smaller, but the rewards tend to be incremental rather than dramatic.

Deductive thinking, by definition, works with what is already known and attempts to arrive at new solutions from a known base. Such thinking is the logical force behind continuous, incremental improvement; it is very valuable when one is searching for more effective and more efficient implementations of the same basic process.

The development of the French video telephone service known as Minitel is a typical example of deductive thinking. Two known technologies — voice transmission and data communications — were integrated into what was thought to be an exciting new application: a combined audio-video workstation. It was to enable users to not only conduct conventional telephone conversations, but also view data files and other information via a small PC-like video screen. For instance, users could make airline reservations, exchange letters and documents, or carry out online shopping. The service, however, suffered from the inherent limitations of its component technologies and was, therefore, not enhanceable to include new functions. With the advent of the Internet and its global reach, endless functional capabilities, and very attractive low cost, Minitel came to be viewed as unexciting, not cost competitive, and technologically dated.

10.4.2 Inductive Thinking

Induction is reasoning from the particular to the general. By observation of particular events, new hypothetical principles are formulated and tested by specific applications. As the results of the tests satisfy the conditions of the hypotheses, laws or results are established. The process of induction attains only a high degree of probability at most and is never considered infallible. Inductive thinking is the thought process of the inventor, the innovator, and the entrepreneur. Recent research has shown that most successful entrepreneurs are very intuitive decision makers — they are able to think "out of the box." Because inductive thinking is not infallible, the inherent risk is higher, but so are the rewards. For instance, strategic thinking is inductive because it takes place in the face of uncertainty. Inductive thinking, for example, could result in application of an existing technology to entirely new and very different uses. Or, it can facilitate the introduction of a new technology to perform functions and services beyond the reach or even possibility of previous ones.

Inductive thinking allows technology to create its own previously unknown uses and, in a way, allows supply to create its own new demand.[5] Just as the digital computer did not simply replace the hand-held or desktop electric calculator and just as the Walkman did not simply replace phonographs and tape recorders, new technology usually creates a demand and market for new uses and applications that had previously not existed. In other words, an important new technology first uncovers the problem and then solves it.

The history of the typewriter is a good example for contrasting deductive and inductive thinking. On the very early models, a manual keyboard moved individual type bars to transfer the writer's thoughts onto the printed page. The next generation of electrically powered, but still manually operated electric typewriters allowed a significant increase in typing speed and the addition of more keyboard functions — significant improvements to essentially the same system. In the IBM Selectric, the individual type bars were replaced by the rotating typing element (dubbed the "golf ball") to increase speed and allow the typewriter to become a fast printing output device.

Today's keyboard-operated desktop or laptop computers offer a quantum leap in functionality, but the keyboard still remains the critical transfer mechanism between thoughts and the spoken word to their symbolic (printed) representation. The emergence of new voice-activated software, however, allows people to dictate to the computer — truly revolutionizing how people are doing their knowledge work.

Throughout the history of the typewriter, deductive thinking has kept all improvements and enhancements of the typewriter still dependent on the manual keyboard. In the end, inductive thinking helped apply a very different and revolutionary technology (speech recognition) to overcome the dependency on the manual keyboard and to enable the translation of spoken voice directly into its symbolic representation.

New technologies free companies from the trade-off between central management control and responsive field sales operations. High-speed, broad-band communications networks allow headquarters to have the same information that sales offices have and to access the data that sales offices do (and vice versa) in real time. With this shared, collaborative capability, every field office and headquarters are part of the same information framework. This means that companies can organize in whatever way they wish (centralized or decentralized), because from an information access and processing standpoint, organizational structure becomes transparent. Modern information technology has enabled many companies to overcome the traditional law of organizational management, namely that centralization and decentralization are mutually exclusive.

10.4.3 The Need Paradox

The miracle of modern marketing is that it persuades people they have needs they do not even know about. Apple convinced people that they needed to have a personal computer at home and at work — and the rest is history. This logic of inductive thinking leads us to the need paradox, which says that it is not always helpful to ask people how they would use a technology to solve their needs. The response will inevitably develop in terms of how that technology might improve something people are already doing. Furthermore, customers are often unclear or uncertain of their own needs, which they cannot clearly define, articulate, or predict. As we will see later, dependence on customer needs alone in determining what the full response to change should be is not only inadequate, but sometimes also risky in terms of continued future adaptability. It is equally important to recognize the rise of an opportunity to offer new technology that, in turn, will create its own new needs. The examples below show how information technology used in innovative ways has helped organizations discard time-honored rules of the hierarchical corporation.

In the hierarchical organization, the traditional decision rule has been that management makes all decisions. This made good sense in a command and control system where management does, in fact, have a broader perspective based on more and better information on reasonably predictable change in a relatively static business environment. In a world of rapid, discontinuous, and unpredictable change, however, the costs of hierarchical decision-making become unaffordable because decisions get made too slowly for the level of responsiveness needed. Today, companies have to push decision-making down and out and empower employees who are actually doing the work or who are in fact closest to the customer to make their own decisions. This is particularly true and needed in *e*-business, where decision-making is distributed along the networked value chain.

Empowerment can only be achieved, however, by giving people not only the authority to make the decisions, but also access to the information that supports it. This is where the integrating technology of decision support systems and collaborative computing comes in. The resulting innovation makes information previously available only to management widely accessible and easy to process with easy-to-use analytical and modeling tools. Individual professionals suddenly have impressive and responsive decision-making capabilities — in the field, in the factories, or in service centers.

In the consulting profession, the traditional operating rule has been that consultants need a physical facility where they can receive, process, store, and transmit customer-related information: the field office. With office and communications equipment installed and an administrative organization in place to support the consultant's activities — from initial client calls through proposal preparation and work delivery — the field office has been the engine room of the consulting business. With the integrating technology of portable, hand-held computers and broad-band, wireless data communications, field employees can receive, use, process, and transmit information almost anywhere without ever having to use physical office space. The subsequent innovation in operating practices has allowed many consulting firms to dramatically reduce their field office infrastructure, transferring the same capability to the consultant's home, or even automobile, by means of the new technology.

One result of portable computing technology being integrated with wireless data communications is that customer service employees of all major car rental agencies now carry small hand-held intelligent terminals. When a customer returns a car to a rental location, an attendant equipped with a tiny hand-held (palm-top) computer meets the driver of the car, retrieves the record of the rental transaction through wireless communication, enters the check-in time and day, fuel status, and the miles driven. The hand-held computer terminal calculates the charges and prints the invoice that is immediately handed to the customer, who never has to visit the rental agency's office.

These examples make it clear that further and even accelerating advances in information technology will help us discard even more traditional rules about how we conduct business — rules that, until recently, appeared inviolate and yet can become obsolete very quickly. The point here is to exploit the potential of information technologies in order to enable the business processes of an organization to be effective and efficient in current terms and, more importantly, to make them highly information sensitive and adaptive to move the organization ahead of its competitors in meeting and responding to change.

This, however, is not a one-time event or something a company does occasionally, say, once every few years. On the contrary, information technology must be incorporated in all business processes and, once this is done, information systems-related activities themselves (such as application software development, customer service, or network management) will become highly innovative and responsive.

10.5 Summary

Achieving adaptability is primarily an information management issue. This means that adaptable business processes in a process-centered enterprise — and the enterprise itself — must be made information intensive. An information-intensive process can, by definition, change as it receives and absorbs new information from its environment. Information management in a process-centered environment means that management can organize resources into tangible capabilities and executable actions by simply manipulating the information representing them.

An important purpose of information technology in a process-centered environment is to provide timely information that helps management determine the appropriate response to unpredictable change. Information technology that translates basic data for useful purposes and, most importantly, for decision-making is a direct contributor to current process performance and adaptability. The most valuable form of technology implementation for this purpose is the networked collaborative system.

The full potential of information technology in enabling and determining process response to change can only be arrived at inductively. Thinking deductively about information technology works with what is already known and attempts to arrive at new solutions from the existing base. It may cause people to overlook its real, long-term potential, what is really important about it, as the nature of deduction encourages them to look at what is, in fact, obvious or trivial.

Part III: Process-Centered Management

11 Basics of Business Process Management

This chapter discusses the basics of business process management — upon which the practical management of the process-centered enterprise is based. It then goes on to examine the relationships and interfaces within individual business processes. The primary focus is on the first phase of business process management: the selection and initial planning of the process.

Given the central role played by business processes, it is ironic that while organizational units (groups, divisions, departments, and sections) are familiar to people in business, processes are not. This is a direct consequence of the command and control approach to management which requires that organizational lines are clearly drawn around specialized functions, denoting reporting relationships; meanwhile, the work process, how work itself is performed, is seldom mentioned and rarely documented.

After almost a decade of worldwide experience with process improvement and reengineering, this is still very much true. The most telling examples are found in the professional services: law firms and consulting organizations. Here, professionals spend enormous amounts of time and effort on developing (in plain English, selling) and delivering business. Yet, only in very few instances can one find any form of formal description of the underlying processes. Many professional firms are collections of solo practices; when a partner or senior consultant leaves, he or she also removes part of the process from the firm without a trace, because the firm never managed to map the partner's approach for getting the work done. Therefore, it should be useful to first examine and review the conceptual underpinnings of process management, to describe how processes are defined, and to suggest techniques for selecting those that should be reengineered.

11.1 Process Management Overview

Over the years, numerous systematic approaches to process management have been proposed, but, beyond statistical process control, relatively few practical solutions to operational process management have been suggested. Most practitioners and reengineering methodologies have focused on process definition and redesign — virtually ignoring the ongoing operation and management of the redesigned process.[1]

11.1.1 The Process Management Roadmap

In the process-centered enterprise, business process management is inclusive: it comprises the planning (definition, mapping, and design or redesign) as well as implementation of individual processes. This is because the prime purpose of process centering is responsiveness to change. Responsiveness, in turn, depends on two key factors: (1) how the process was designed, and (2) how it is operated. What takes place in the real-life process, each step of the way, to make it adapt to unexpected change?

In fact, this book defines business process management in terms of three consecutive and chronological phases. These are the phases individual process teams will implement to reengineer their business processes for adaptability and to align them to create the process-centered enterprise:

1. *Phase One.* Process planning — the definition, mapping, analysis, and design of the process
2. *Phase Two.* Process transfer (also known as process launch) — implementation planning and process deployment
3. *Phase Three.* Process operations — the day-to-day execution of the newly designed (reengineered or redesigned) process

Once several individual processes have been designed or reengineered, process centering can be addressed, provided the process design included structural features and human attributes specifically aimed at process adaptability. (The specifics of this design approach will be discussed in Chapters 15 and 16.) This proviso is important because traditional reengineering methods have focused on achieving improvements in current process performance only. Accordingly, the alignment of the enterprise around its key processes, newly designed for adaptability, can now be completed by combining multiple individual process management efforts into an enterprise-wide approach. The result is clear: successful process-centered enterprises are managed by managing their business processes.

This approach to business process management offers some powerful opportunities to leverage the performance of individual business processes:

- Process management provides the best platform from which to plan and directly link managerial actions (decisions and commitments) to procedural work activities within the process. As we will see later, holistic process

Table 11.1. Major IBM Corporate Business Processes

Process Ownership	*Business Process*
Worldwide ownership	Market selection Customer requirements Hardware development Software development Production
Coordinated ownership	Market information capture Marketing Financial analysis (pricing) Plan integration Customer fulfillment Customer feedback
Shared ownership	Services development Solution integration Service Customer relationship

Source: IBM Corporation (Armonk, NY, 1994).

management is based on understanding the relationship between managerial decisions and commitments, on the one hand, and the resulting work activities, on the other. From this viewpoint, managerial actions become an ordered, inherent part of business process management — something the traditional input-process-output (IPO) process model had difficulty resolving.

- Process management presents a continuum for turning the adaptability of individual processes into an effective, enterprise-wide response to change. As discussed earlier, adaptability is primarily a matter of process design. However, translating it into real-life, real-time response activities is very much a challenge for operational process management — that is what process owners and process management teams do day after day. They do it by systematically linking the execution of procedures within a process to the organization's strategies and overall business performance. This is possible because process management treats the enterprise as a system whose parts contribute to the success of the whole. An inclusive perspective allows management to focus on both the whole and, at the same time, understand all the parts that make it succeed.

11.1.2 Classification of Business Processes

Most companies, even the largest and most complex ones, can be broken down into no more than twenty major business processes. Table 11.1 shows the list of IBM's 15

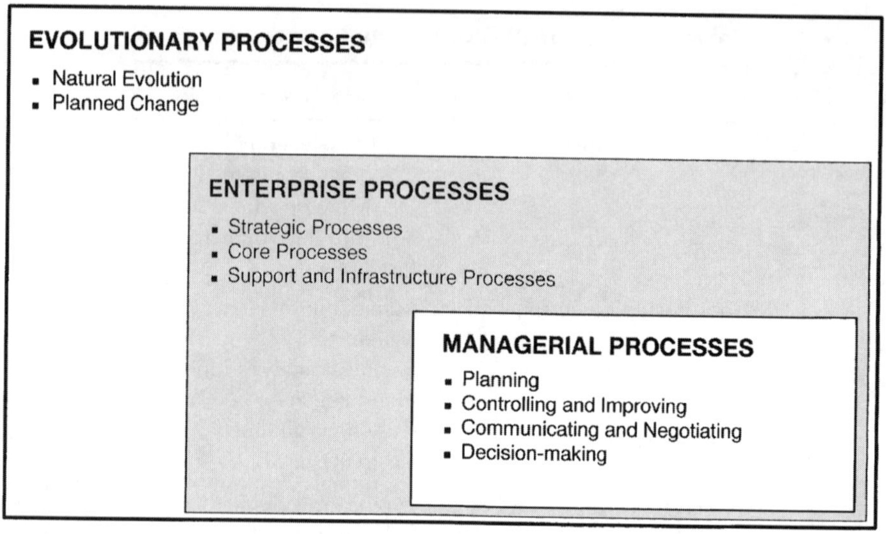

Figure 11.1. Taxonomy of Business Processes

major corporate business processes, subdivided into three categories of ownership. Xerox Corporation is said to have 14, and Dow Chemical is said to have nine principal business processes.[2] The ordering, categorization, and designation of business processes is as varied as there are authors on the subject.[3,4,5] This book has chosen to use three, rather than a score of, major categories that are generally recognized and implemented by leading practitioners of business process management (Figure 11.1):

1. *Evolutionary processes:* Sequence of events over longer periods of time, often pre-ordained and irreversible. Most of these processes occur in nature, the economy, or in the life of societies and civilizations. Evolutionary processes tend to provide the envelope for a large number of processes belonging to the other categories.[6] Relatively few evolutionary processes take place in companies or other similar organizations. Organizational learning is one important example of evolutionary processes within an organization, where an autonomous organizational trend is influenced by direct human intervention. An example would be augmenting the on-the-job experience of process teams with special training in team building, the use of collaborative groupware, and so on.
2. *Enterprise processes:* Business processes that represent all the work performed within an organization to meet external or internal requirements (see Figure 2.3 [????????]). Their overriding purpose is the same as one of the enterprise's: value creation or contribution to it. Divided into broad groupings by purpose, this category can be further subdivided into:
   a. *Strategic processes,* which are necessary to identify and manage the enterprise. They deal mostly with enterprise-wide, strategic issues, and their outcomes usually have long-range implications. Included here are

governance and business planning, the process of managing business processes, managing information technology, finance, and the like.

   b. *Core processes,* which are essential to enterprise effectiveness. They are the processes that directly interface with customers and generate customer value. This grouping includes developing, producing, and distributing products; sales and service; order fulfillment, and the like.
   c. *Support and infrastructure processes,* which are necessary for sustaining the day-to-day functioning of the enterprise and the ongoing operation of its strategic and core processes. Included here are such processes as human resources, research, procurement, legal services, facilities management and maintenance, and the like.

3. *Managerial processes:* sequences of information and knowledge-intensive activities that are usually aimed at producing such management outcomes as agreements, decisions, and commitments. These outcomes establish the preconditions for value creation and are then used to guide and control the implementation of procedural work across all of the enterprise. For this reason, managerial processes that are appropriately embedded in the enterprise processes include:
   a. Planning
   b. Controlling and improving
   c. Communicating and negotiating
   d. Decision-making

These managerial processes, as related to process-centered management, will be further discussed in the next chapter.

11.2 Process Planning

Process management is process-specific in that it systematically defines, maps, analyzes, and redesigns individual business processes in response to changes in the environment or in support of opportunities for generating new demand and creating new requirements. For the effective functioning of a business process, management must clearly define and be able to rely on a number of fundamental factors which are always current for that process. To illustrate these, we will use Union Camp's Fine Paper Division throughout this chapter.

- *Overall, unique purpose (mission) of the process:* Common sense and business prudence dictate that there should not be multiple, competing business processes in the same company producing the same or competing output for the same customers. Nevertheless, in real life, this often happens, either by design or, more likely, by oversight. (Examples of this include multiple business units of the same corporation competing for the same customer — as, for example, the Fine Paper Division's two mills trying to produce the same product to satisfy the same customer order.) By establishing a unique purpose, formalized in the mission statement for each and every business process within the enterprise, wasteful duplication can be avoided.

- *Process performance goals:* Process goals are primarily represented by the set of committed process and subprocess outcomes. For example, Union Camp's order-fulfillment process is committed to providing a complete answer to routine customer inquiries on the first call. This, then, is one of several ongoing performance goals for the entire process to adhere to.
- *Process boundaries* (interfaces with customers, suppliers, and other processes): Union Camp's Fine Paper Division serves three major market segments: converters, merchants, and retailers (Staples and Office Depot, among others). For any given order, the customer interface and certain procedures within the order-fulfillment process change so that they respond effectively to the market segment with which the customer is identified.
- *A defining structure* (made up of information and work flow, measurements, and controls): A source of customer needs, the customer interface will also indicate the conditions of satisfaction to be met by the deliverables. For instance, different customers may have different quality specifications, which, in turn, will determine the set of measurements to be used in the course of satisfying the specific order.
- *Status of commitments:* While the minimum acceptable rate of on-time deliveries is 99% for the order-fulfillment process over a given period of time, certain customers or competitive situations may dictate an even higher standard. Several major merchants in the fine paper market are now requiring *Six Sigma* performance when it comes to deliveries (which is no more than 3.4 failures in a million cases).
- *Current performance of the process* (effectiveness and efficiency): Actual performance against any of the process goals, response time to inquiries, on-time delivery, and paper quality must be constantly monitored so that this information is available on a continuing basis to customer service representatives, the process management team, and the process owner.
- *Deployment of process capabilities* (work activities and resources, such as people, materials, energy, equipment, and information, being used to produce the desired outcome): In order fulfillment, one key capability is transportation, the various carriers that perform the actual product delivery to the customer's warehouse or office location. Depending on order size, destination, cost, and several other factors, the Union Camp's carriers are chosen order by order, a good example for constantly changing process capabilities.

Through the appropriate use of this information, successful process management will be able to ensure process adaptability in an uncertain environment and on an ongoing basis, as manifested by one or more of these results:

- *Dynamic change in the status of commitments with no ambiguity about related outcomes or assigned accountabilities:* Fine paper shipment destinations are often unexpectedly changed from one warehouse location to another by

Union Camp's customers. The order-fulfillment process must have a record of the original commitment and who made that commitment, the person accountable for the delivery. The new shipment destination is a new commitment; however, accountability for the outcome may stay with the same individual in the transportation subprocess.

- *Ongoing, continuous coordination*[7] *of commitments and the management of related outcomes to ensure that process performance goals are met:* In order fulfillment, there is ongoing coordination between two major sources of commitment in satisfying a given customer order: (1) new paper production in the mill, and (2) current inventory of finished product. A given order may be satisfied from one or the other, or both sources, dependent on a number of conditions. Coordination between new production (commitment #1) and delivery from inventory (commitment #2) is one of the ongoing tasks of order-fulfillment management.
- *Pre-determined and* ad hoc *process capabilities:* Many times, there is no pre-planned procedure and, therefore, no available resource to face a new, unexpected situation. In such cases, outsourcing may be the best solution to create the necessary capability. One capability that is almost exclusively outsourced is transportation.
- *Dynamic change in process boundaries (redefinition of customer-supplier relationships) and structure (process reconfiguration or redesign):* One example of a change in process boundaries would be for the Fine Paper Division to create its own transportation fleet and no longer rely on outsourcing transportation. This would probably also require a partial redesign of the order-fulfillment process itself.
- *Earliest possible detection and correction of poor process performance:* Detection of paper quality problems in new production (a situation known as lower grade output) is extremely important, because a given set of orders may have to be filled, without any delay in delivery, from another mill or from inventory. Also, the production process may require significant improvement to prevent the recurrence of the quality problem.
- *Full compatibility between process performance goals and the work performed (people's roles and responsibilities derived directly from the process flow):* In Union Camp's order-fulfillment process, not only are process professionals fully aware of the overall process performance goals (e.g., 2-minute response time, and so on), but they can also tell how their individual performance objectives contribute to that overall goal.

In the next sections, we will examine issues related to the clear and unambiguous definition of the business process in terms of purpose, goals, boundaries, and structure.

11.2.1 Process Identification and Mapping

In any company, business processes are the natural sequence of business activities. In a functionally organized company, however, they are often fragmented and obscured by functional specialization and the related organizational structures. In the paper industry, for example, the order-fulfillment process is critical to customer satisfaction and competitive position. Yet, in many paper companies, order fulfillment is visible only through the functional departments where pieces of work related to order fulfillment are done: paper production, customer service, sales, transportation, warehousing, accounts receivable, and so on.

In light of this, the first task usually assigned by enterprise management is the identification of all business processes. As this is done, process naming becomes the next important task. One way to better identify the processes that make up an organization is to give them names that differentiate them from organizational units: divisions, departments, and sections. Process names should express the purpose, work product, and, at times, the work content of the business process. Such names may also imply the beginning and end states between which work takes place. In the author's practice, the preferred approach to naming processes is the *verb plus noun* formula that will produce names such as:

- Develop strategy
- Evaluate customer request
- Negotiate commitment
- Develop product
- Procure supplies
- Manufacture parts
- Assemble product
- Fulfill orders

Just as functionally organized companies have had their organization charts, process-centered enterprises have process maps that give a clear picture of how work and information flow through the organization. Organization charts, however, are designed to show hierarchical relationships of command, control, and reporting; they group people by the functional similarity of the work they do. The sales department, where all people do sales-related work, is shown on one chart; the finance department, where people do financial work, again has its own chart; and so forth.

Process maps, on the other hand, show the natural flow of all types of work and information across the organization, from the supplier to the customer, related to the same outcome and as they exist in reality. Accordingly, process maps group people together by the outcome of the work they do. A process map also creates the terminology that will help people and, especially management, practice the management of business processes.

Process maps resemble the flow diagrams used for decades by systems analysts and software programmers. In this book, we have used three conventions for process mapping, each serving a particular purpose in representation. This is what the maps look like:

1. Enterprise-level process maps, using process icons to denote major, corporate-wide business processes (Figure 11.2.a), offer a "view from 100,000 feet"

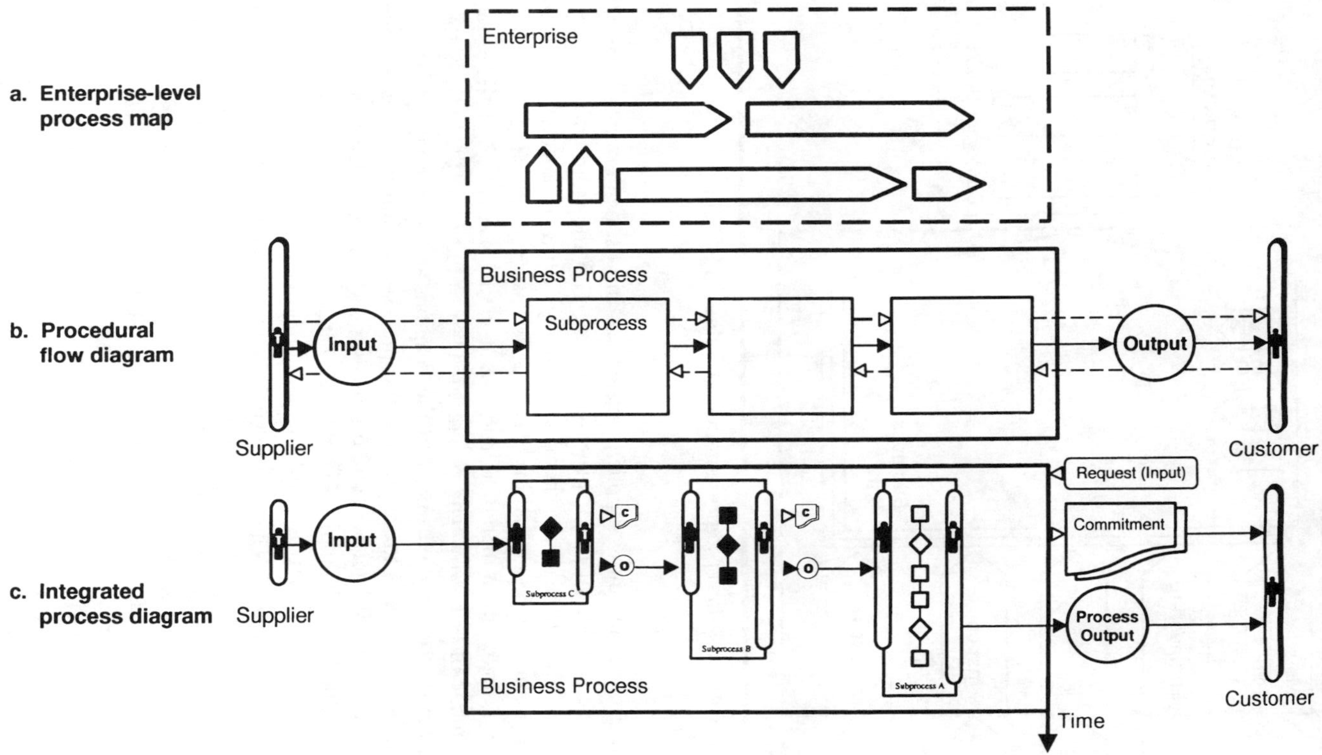

Figure 11.2. Process Mapping Conventions

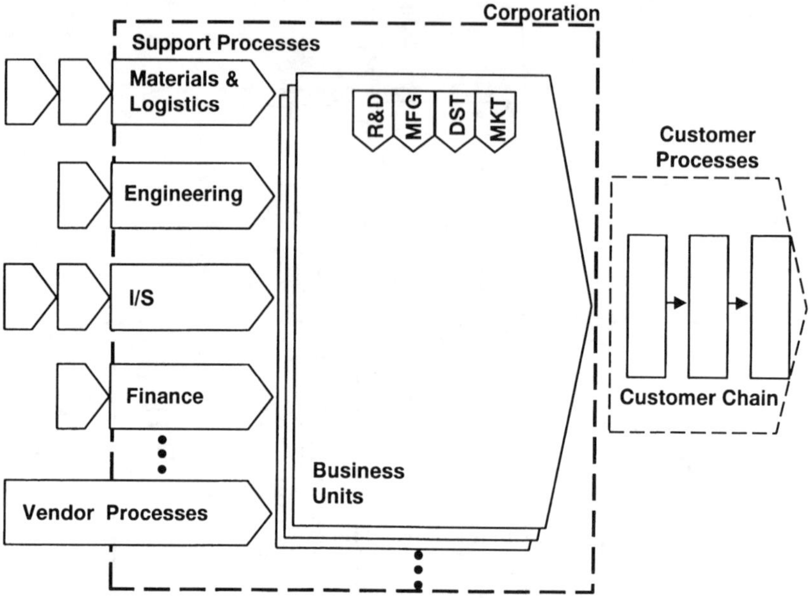

Figure 11.3. Enterprise-Level Representation of Value-Add Chain

which does not show suppliers, customers, inputs, or outputs — just the relative position of major processes. Figure 11.3 shows an enterprise-level process map of a major chemical company. Using an analogy from geographic mapping, we can think of enterprise-level process maps as the equivalent of the political map of a major country, such as the U.S. or the Federal Republic of Germany. The geographic map would show the various states within the country, plus perhaps their capitals and a few major rivers. The process map shows the major business processes within the enterprise with perhaps the key interfaces and outputs for each process identified.

- Procedural flow diagrams (Figure 11.2.b) are primarily used to map discrete, procedural work activities and information flows (essentially the IPO process model). Figure 11.4 is a high-level process flow diagram depicting the corporate information services process of a major pharmaceutical company. The geographic analogy for the procedural flow diagram would be the political map of a single state, showing the political subdivisions, such as counties, with county seats and other major cities and towns identified along with a few railway lines and major highways.
- Integrated process diagrams (Figure 11.2.c) represent the holistic process model (discussed in Chapter 13) of customer-supplier relationships integrated with the procedural work flow. Such diagrams may also include a time axis, parallel to the procedural flow and perpendicular to the chain of customer-supplier relationships. Figure 11.5 shows the integrated process

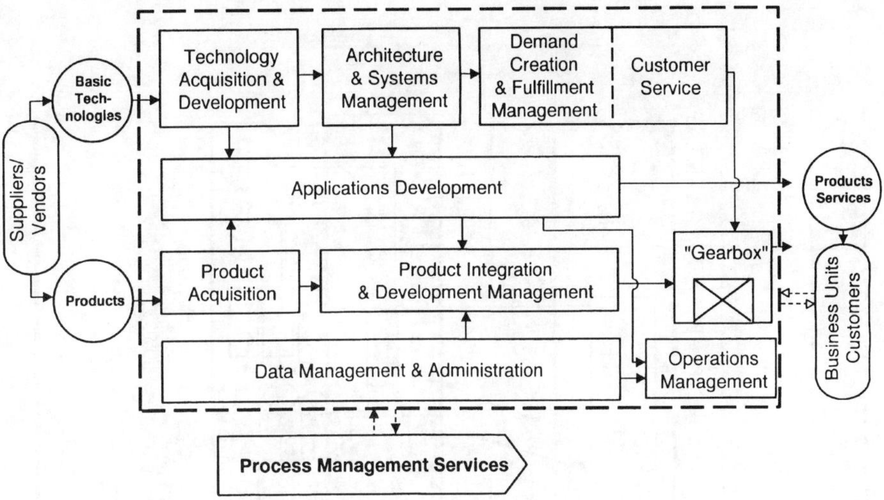

Figure 11.4. Corporate Information Services Process

diagram for the forecast reconciliation process of Union Camp's Fine Paper Division. Note that the diagram is rotated 90 degrees counterclockwise for easier reading; the time axis here is horizontal. The process depicted in Figure 11.5 shows three commitment dialogues (shaded areas) — one prime commitment (C2c) and up to 22 possible subordinate commitments — needed to produce the outcome: the delivery of the "tons to ship" forecast.

The analogy here would be the geo-economic map of a single state, showing no political subdivisions but all major population centers, all geographic and functional features such as all rivers, the entire network of railways and highways, all important airports, seaports, etc. In other words, it shows the major features that connect the various business and population centers to allow them to carry out trade, communications, travel, and other activities essential for the economic life of the state. All of these process maps have two important characteristics: (1) simplicity, and (2) consistent orientation of key activities to the customer by showing them converging on the output of the process.

11.2.2 Process Selection for Reengineering

Once the business processes of an enterprise have been identified, named, and mapped at a detail level that shows the process itself and its six to ten key subprocesses, the next step is to select the one or several business processes that require changing and perhaps radical improvement, in terms of current performance and adaptability. Selection is a job normally reserved for senior management (as represented by the transformation council). It is based on a sequence of economic value

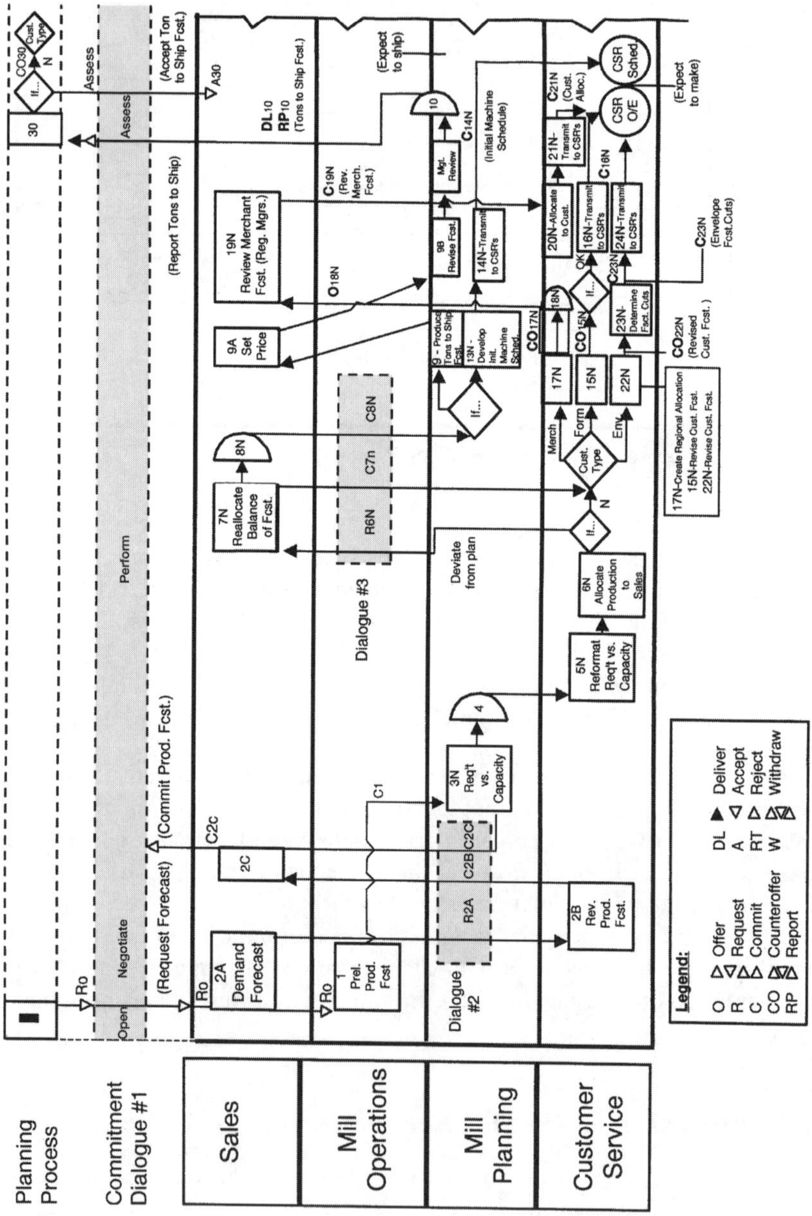

Figure 11.5. Reconcile Forecast and Balance Machine Schedules

added (EVA) and intellectual value added (IVA) studies (see Chapters 3 and 4). These studies are performed on processes around which the alignment of the enterprise is planned.

Once a business process has been identified and selected, a process owner is named and a business process team is assigned. These actions are strategic and, therefore, should be undertaken by senior management. In many companies, it has also been accepted practice to provide an initial process *mission statement* and a set of initial process performance goals to the process owner as overall guides for the management of their particular business process. This is strictly senior management's initial view of what is expected of the process within the enterprise framework. It will be among the early tasks of those named responsible for managing the process to validate.

The mission statement is an expression of the purpose of the business process, its *raison d'être.* As such, it should be specific for the given process, a manifestation of the concept of uniqueness of purpose clearly indicating that there exists no other such process within the enterprise. The process mission statement for Union Camp's Fine Paper Division's order-fulfillment process states: "The Fine Paper Division's Order Fulfillment Process economically and efficiently provides [fine paper] goods and services to meet customer needs on a timely basis."

Process *goals* are quantified indicators of performance that the process is expected to attain within a defined time frame. The process performance goals for the Fine Paper Division's order-fulfillment process included:

- Achieve 99% on-time delivery.
- Respond to all routine customer inquiries on the first call.[8]
- Reduce cycle times by at least by 20%.
- Reduce inventory levels by 25% (average through the business cycle).
- Reduce delivery expenses by $3 million.

The process owner and his/her process team, the process mission, and performance goals are also referred to as the key elements of the process management infrastructure.

11.2.3 Process Definition

At this point, senior management's active involvement ceases; the newly appointed process owner and process team take over. The next step of process management is then undertaken by the process owner and the team to define the business process in its *current state* and to validate its mission and initial performance goals. The sequence is important: one cannot validate the mission and goals of something without first understanding what that something is. Process definition, in this sense, is the systematic and methodical understanding of the existing business process. A given business process can be defined and, therefore, understood only in terms of the following features and attributes:

- Mission and performance goals (the latter representing the initial set of required commitments and outcomes)
- Management infrastructure (process owner, process management team, and stakeholders)
- Network of customer-supplier relationships
- Procedural flow (work flow and information flow)
- Process resources
- Process measurements and controls

Whether the goal is to improve the existing process or replace it with an entirely new design, one overriding objective must always be included: to create a superior process design in terms of *adaptability*. For this reason, the process team does not need to analyze and document the existing business process to great levels of detail. Rather, the team requires a level of definition that is sufficient enough to provide insight into the current performance of the process.

The decision to redesign or reengineer the process cannot be made without understanding its current performance. One of the most frequently committed errors in the early phases of process management is to try to define and then analyze the business process in excruciating detail (the proverbial "paralysis by analysis") rather than focusing on understanding and evaluating its current performance. The business process team must determine what the output of the existing process is, what the customer of the process does with that output, and whether that output adds value to the customer's own processes and operations. This is the real indicator of process performance.

11.2.4 Customer Requirements

The best place for the business process management team to start with the structural definition of a business process is at the customer end: with the output the process is producing for its customer. Process output is usually the only unambiguous fact known about an existing, but not formally defined, business process. The adage of "the process is defined by its output" is not far from the truth; once process outputs are listed, it is relatively simple and straightforward to identify and logically link all subprocesses and activities necessary to produce a given output. The linking is usually done in reverse sequence; that is, from result to cause, from output to the activities producing it. This is also known as mapping the process at a detailed level. In process reengineering, this approach is known as the right-to-left rule of mapping, because in a conventional procedural flow diagram, one is proceeding from the righthand, output side leftward to the input side (Figure 11.6).

Mapping process output against customer needs and expectations is the prerequisite step to the analysis of current process performance in terms of effectiveness and efficiency. The conclusions drawn will form the essential guidance to the business process management team in its decision as to whether their business process needs major reengineering, only incremental improvement, or no change at all.

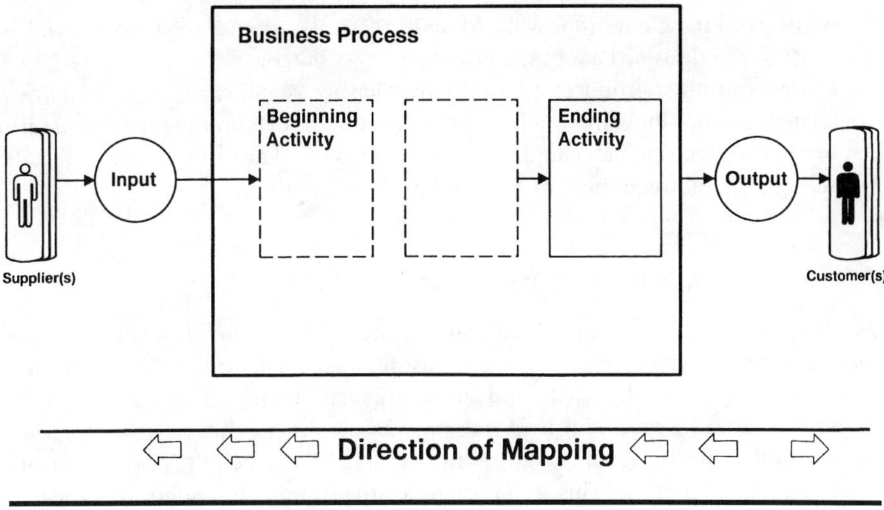

Figure 11.6. Right-to-Left Rule of Process Mapping

Starting process definition at the customer end is consistent with the basic premise of total quality management, which is essentially a management approach with a customer orientation. The comprehensive identification of the customers of the business process and their requirements is one of several organic links between total quality management and business process management. (Another link is the commonality of some of the management tools used; the third is the integration of continuous quality improvement into the operational management of business processes.)

As we have said before, a business process is not merely a series of work activities that in the aggregate, make up the workflow. Rather, the process should be viewed as a series of customer-supplier relationships which include the communication of requirements, the negotiation of commitments, and the management of the work flow governed by them. It is, therefore, essential that in the definition of the business process, its customers — both internal and external to the process — are identified and their needs and expectations determined.

The basic questions here include:

- Who are the customers?
- What are their real needs?
- What are the benefits they expect to derive when those needs are met?
- What are their problems if their needs are not met?
- What problems do they currently have with the products and services provided by the business process?
- What are the resulting real requirements?
- What customer processes require the outputs of the business process as input?

The answers to these questions will add up to the real customer requirement which is crucial to the decisions about the desired process outcome.

Process output is produced to meet customer needs and expectations. The two are related, but not the same; yet both are important in determining the customer's requirements against which commitments can be made. The distinction to be made between needs and requirements is, therefore, very important.

11.2.4.1 Customer Needs and Expectations

Need represents a lack of something requisite, desirable, or useful; it is a condition requiring provision or relief.[9] It is an existing state that can be defined with a reasonable degree of objectivity and precision, such as hunger or cold, need for transportation, or a specific type of training. On the other hand, *expectation* means anticipation of the future state of things and, especially, of benefits to be derived from satisfying a known need. (This is important for the supplier to understand, because the benefits expected by the customer will provide a good indication of the value perceived.)

Given a recognized need, the related expectation will have two components: the manner in which the need is to be met and the anticipated benefits to be derived. Meeting a need without regard to the expected manner is only the necessary, but not sufficient, condition of customer satisfaction, because not all the anticipated benefits may be realized from just any manner in which the need was met.

11.2.4.2 Customer Requirements

The distinction between need and expectation is critical to formulating the following definition: *A requirement is a formal statement of a need and the expected manner in which it should be met.* As we will see in Chapter 13, requirements (also called *requests*) open the formal dialogue between customer and supplier, eventually leading to a commitment by the supplier to produce a desired outcome to satisfy that request (or a negotiated version of it). Thus, the supplier's commitment always includes the agreed-to *conditions of satisfaction* to be met by the process output. Once these conditions are met, the customer's obligation is to accept the output and compensate the supplier for it in the agreed-to manner.

We can conclude that, rather than analyzing a process to death, a comprehensive understanding of the customers, their goals, expectations, and potential problems is key to an effective customer-supplier relationship which goes far beyond the input-output mechanics of the business process. To this end, the business process management team may, by design, include key customers of the process. In addition, the entire team, or at least some of its members, may make collaborative arrangements with other key customers that will allow them to gather critical customer-related information by means of observation, interviews, and even participation in key customer business processes.

11.3 Summary

The three fundamental categories of processes include evolutionary, enterprise, and managerial processes. It is the last two categories that we regularly encounter both in private sector enterprises and public sector organizations.

In the process-centered enterprise, business process management is inclusive: it comprises the planning (definition, mapping, and design or redesign), as well as implementation of, individual processes. This is because the prime purpose of process centering is responsiveness to change.

This book defines business process management in terms of three consecutive and chronological phases. These are the phases individual process teams will implement to reengineer their business processes for adaptability and to align them to create the process-centered enterprise: (1) process planning, (2) process transfer, and (3) process operations.

Preceding the actual start of the process planning phase, senior management identifies and selects business processes intended for reengineering. The process management team, led by the process owner, first validates the process mission and performance goals. Then it proceeds to map the existing process, identify its customers and their needs, and establish the chain of customer-supplier relationships and related procedural activities, which represent the basic structure of the process.

One of the central tasks here is to determine the needs and expectations of these customers, which results in formal customer requirement statements. It is these requirements that will become the subject of negotiated commitments, and the drivers of all subsequent process activities.

12 Effective Process-Centered Management

This chapter continues the discussion on business process management by examining process performance, addressing commitment coordination and adaptability, and, finally, introducing process alignment.

12.1 The Operational View

Operational, process-centered management must provide the organization with integrated, continuous control of all key activities required to fulfill agreed-upon conditions of satisfaction. This is true whether the deliverable is a physical product, function, service, reliability, or any other major component of a customer's requirement. Because the performance of any key activity in a business process can substantially affect overall enterprise performance, the key to modern process management can be stated as follows:

> *The business process must be designed and operated so that the interrelationships of all key assets (people, equipment, materials, energy, information, and knowledge) and the activities they perform must result in effectiveness and efficiency at any given time and produce a positive cumulative impact on total business performance.*

12.1.1 The Two Dimensions of Process Management

Most executives are still unaware of the powerful implications of the systemic properties of processes and the leverage they can gain from the consistent implementation of process-centered management. Arguably the most important among these implications is that the performance of a process — an open system — is not the sum total of the performances of its component parts (subprocesses and activities) taken separately. The product of their interactions, *synergism,* explains why the performance of a process can be greater than the sum of the contributions of its subprocesses.[1]

Functional organization and conventional managerial approaches, including continuous improvement and traditional process reengineering, focus on the actions of the parts rather than their interactions. Most methodologies provide a rich toolkit for identifying, analyzing, and then improving the functional units, subprocesses, activities, and tasks. It is assumed, then, that if each part is made to work as required when taken separately the entire business process will, also. This, however, is not always the case and is certainly not a correct assumption to make.

Accordingly, effective process management must focus on the interactions of the process parts: the series of commitments and communications that drive the actions of those parts. These commitments and communications, in fact, are the products of the value-chain of customer-supplier relationships that make up the business process.

The two dimensions of operational process management, then, are (1) managing commitments, and (2) managing actions (work activities).[2] The distinction between the two is critical. The management of commitments requires the empowerment of those responsible for subprocess outcomes to (a) make or change commitments, and (b) decide what capabilities[3] to put in place for delivering on any given commitment.

12.1.1.1 Managing Commitments

Only trained and empowered people have the ability to handle the unpredictable and unexpected. A change of sufficient magnitude in customer requirements necessitates a decision to change one or more existing commitments or establish new ones. Changed or new commitments mean new outcomes. This is why, in the face of an increasingly unpredictable business environment, considering the commitment dimension of a business process becomes essential.

The adaptability of the process and responsiveness of process managers are maintained by keeping track of the commitments made between empowered parties in each customer-supplier relationship. The ordered set of linked commitments related to a given process outcome is known as the *backbone network of commitments.* Each commitment in the network has associated with it a subprocess outcome and related conditions of satisfaction which have been negotiated case by case between the two parties concerned. The totality of subprocess outcomes will lead to the desired overall process outcome. Inherent in this concept is the variability of the network: different process outcomes may have different commitment configurations associated with them. This leads to the logical conclusion that commitment network reconfiguration is another necessary prerequisite of process adaptability. This backbone concept of commitment management is illustrated in Figure 12.1. In the network configuration shown, subprocesses A, B, C, and D are linked to produce an already committed output. The output may have to be modified to satisfy a new

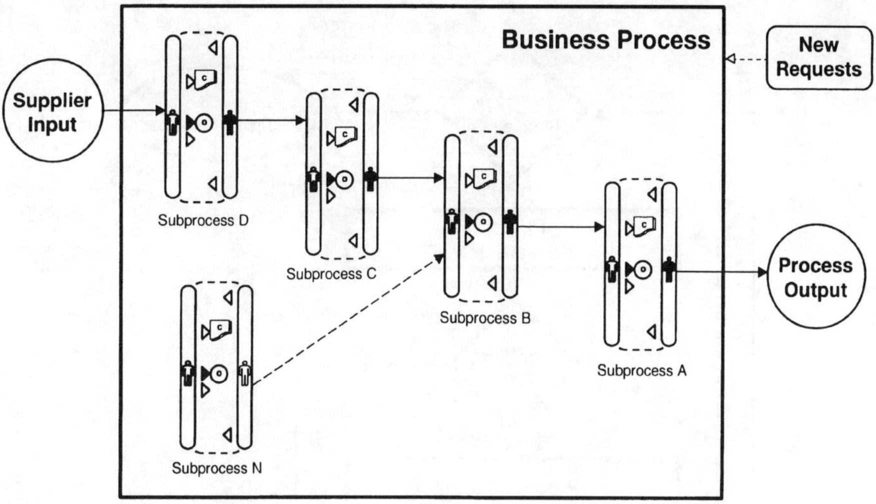

Figure 12.1. Backbone Network of Commitment Management

incoming request. This could necessitate, for instance, a supplementary commitment from subprocess N as well.

The backbone concept comes into play every time someone is wait-listed for an airplane seat, a hotel room, or anything that requires advance reservation and, consequently, depends on a commitment. The outcome now becomes dependent on someone other than the person in charge of reservations; the dependency and commitment are transferred to the person responsible for the waiting list. Once the list is exhausted, the making of reservation commitments returns to the reservation agent.

12.1.1.2 Managing Actions

Managing actions means the ongoing coordination, measurement, and control of procedural sequences (subprocesses) and their outcomes. Of course, each subprocess outcome can, in turn, be broken down into even smaller activity and task-level component results. The breaking down of a business process into its smaller component parts is also known as *decomposition*. Deciding to how many levels this should be carried out is a matter of management efficiency and employee empowerment. From a practical standpoint, the lowest level of decomposition should coincide with the financial or resource-related decision-making authority of individual employees in the process.

By definition, procedures deal exclusively with situations that have been thought of in advance and for which a predetermined sequence of actions can be devised;

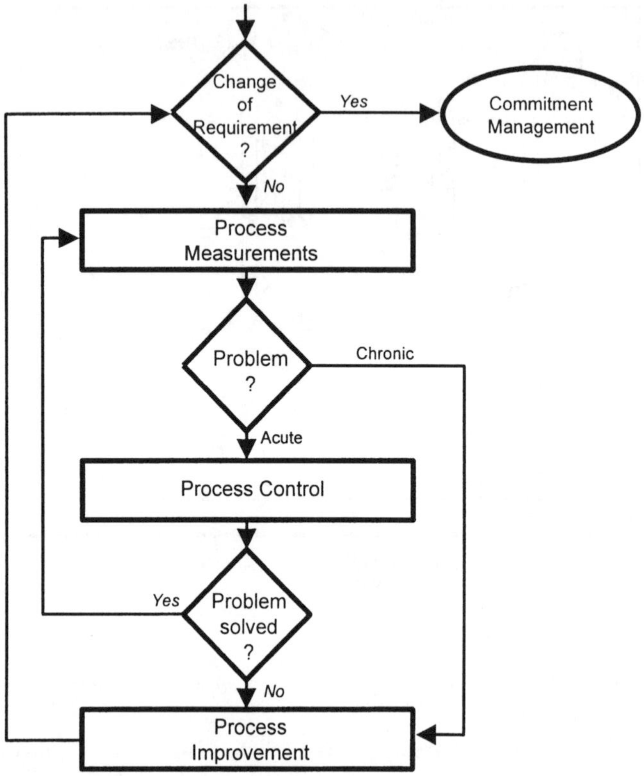

Figure 12.2. Operational Process Management

hence, they are predictable. Once fully defined and designed (or redesigned), managing the procedural dimension of a business process is essentially that of managing the workflow and the movement of interim products from subprocess to subprocess, activity to activity, always following the chain of customer-supplier relationships defined by the backbone network. A manufacturing line that is making a part for a customer according to a pre-determined design is an example of this. The day-to-day management of workflow uses performance measurements to monitor process performance in terms of effectiveness and efficiency.

The fundamental ingredients of operational process management include (Figure 12.2):

- *Process measurements:* Variation in the output of every subprocess and sometimes even certain activities must be measured and identified so actions can be taken to ensure conformance with the agreed-to conditions of satisfaction as they relate to the output of the process.

- *Process control:* The managerial process of eliminating acute problems and ensuring that the process is stable and its output predictable so that it can continue to meet committed conditions of satisfaction.
- *Continuous process improvement:* The managerial process of eliminating chronic problems and ensuring that ongoing improvement becomes permanently embedded in the process. As a result, process output will always meet the already agreed-to conditions of satisfaction. Continuous improvement of the process is one precondition of process adaptability.

12.1.2 Commitment Management and Communications

Business processes include sets of coordinated activities performed by resource-consuming components, among which people are the most important. People are not merely a process resource, as the more traditional, mechanical input-process-output (IPO) process model would have them be, but are the most important asset because they are the intellectual value added part of the business process. By their timely actions, they ultimately ensure that processes adapt and respond to change when needed.

Research by Flores and others[4] resulted in the identification of a powerful set of tools for business process transformation. These tools are based on the argument that the essence of process management is coordination through language, carried out by empowered and accountable individuals (starting with the process owner) in a highly disciplined and organized manner. Scherr[5] and Haeckel[6] argue further that the essential content of this language is the negotiation and agreement over commitments that then drive all process activities.

The chain of customer-supplier relationships represents the essential structure of the process in terms of commitments for subprocess outcomes. Coordination within the business process is the totality of customer-supplier communications in support of the commitments made to the external customer. Commitments made among internal customers and suppliers of the process are, in turn, based on a set of decisions which ultimately address the question of which capabilities should be generated within the process and which should be outsourced to outside suppliers. Performing process work by our own resources means incurring further communication and process costs. Outsourcing the same work to suppliers of the business process will lead to added transaction costs.

Transaction-cost economics is essentially the management of trade-off between internal processing and external outsourcing, on the basis of costs and economic value. The ability to switch between internal processing and outsourcing is essential to the adaptability and responsiveness of the business process to change.[7] Many businesses have backup plans for emergency production situations. Those plans include outsourcing critical capabilities which must be available on a continuing basis. A dozen ways to get something done is not an uncommon strategy in organizations that have become experts in process management.

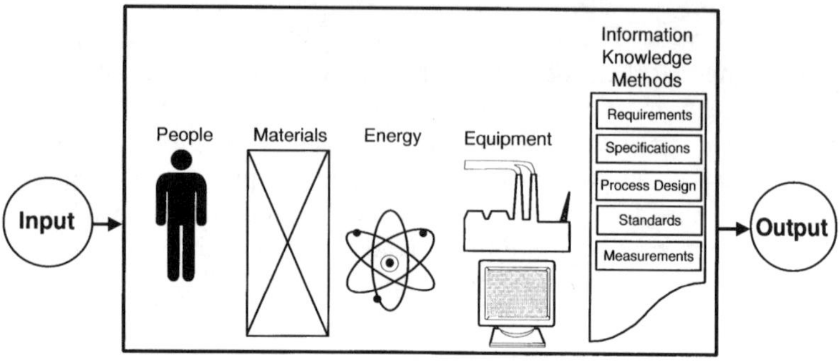

Figure 12.3. Business Process Resources

continuing basis. A dozen ways to get something done is not an uncommon strategy in organizations that have become experts in process management.

It has been estimated that if current trends continue, over two thirds of the workforce will be contract workers by the year 2020. In process terminology, this means that most of the procedural work will be outsourced so the few full-time employees hired by organizations will do knowledge work, communications, coordination of the outsourced work, and activities that involve decision-making, administration, and management. Actual production of items or delivery of services may be left to the outsourcing vendor.

12.1.3 Process Resources

The operating process consumes quantifiable amounts of mostly renewable[8] resources in producing its output. While it is sometimes convenient to align these resources with the two dimensions of process management — managing commitments and managing the procedural work activities — they will ultimately fall into five asset categories (Figure 12.3):

1. *People* (human resources) who do all of the managerial activities and some of the procedural tasks
2. *Materials* used by people and equipment in performing their specific activities (which includes raw materials to be processed, catalysts, chemical agents, and office supplies, just to name a few)
3. *Energy* consumed by both people and equipment in performing work
4. *Equipment*, including all machinery, information technology and communications hardware, tools and devices (including pens and pencils), physical plant, buildings, and facilities. (This category of resources includes most energy- and material-based capital assets assigned to a specific process.)
5. *Information and knowledge* (including methods) used, processed, or produced by the procedural activities of the process, as well as the information

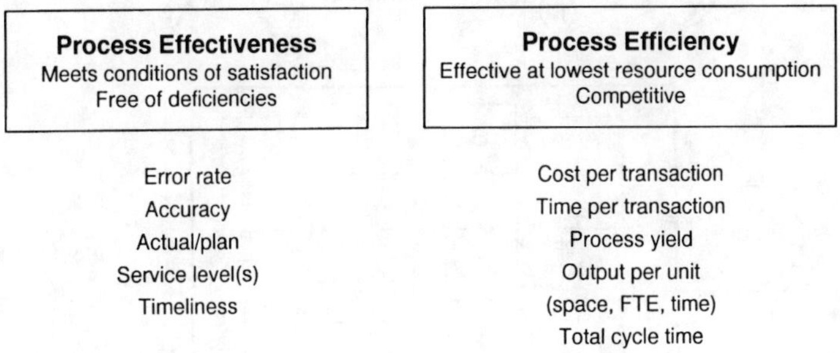

Figure 12.4. Business Process Measurements

The rate at which the business process is consuming these resources while producing output that meets the agreed-to conditions of satisfaction is the measure of its resource efficiency. Because most, but not always all, resources can be measured in terms of cost, it is convenient to express resource efficiency also in terms of cost or measures related to cost.

Cost efficiency of a given process may be expressed in terms of its *period cost* (annual cost, hourly cost, etc.) or *cost ratios* such as cost per unit output, cost per transaction, and so on. As discussed in Chapter 3, an important efficiency-related cost measure is the total cost of quality, or the cost of nonconformance (cost of poor quality).

Performing process work also takes *time.* While, technically speaking, time is not a resource because it is neither renewable nor is its availability process specific, the amount of time (known as cycle time) required by individual process activities or the overall process to produce their output is an indicator of time efficiency. (See Chapter 16 for a more detailed discussion of cycle time and its reduction to enhance process responsiveness.)

12.1.4 Process Measurements and Controls

Establishing and using the correct measures are crucial to the success of managing business process performance. Appropriate measurements provide the basis for determining how well (effectively) and by consuming how much resource and time (efficiently) customer requirements are met (Figure 12.4).

The choice of process measurements is also critical for determining the value contribution of an individual process to overall enterprise performance. In a process-centered enterprise, individual process performance must be linked to overall business indicators, such as product cost, sales, revenue, profit, return on investment (ROI), productivity, and the like. In the long run, one overall goal of managing business process performance may be to ascertain if all key business processes are measurably "best in class" or "world class."

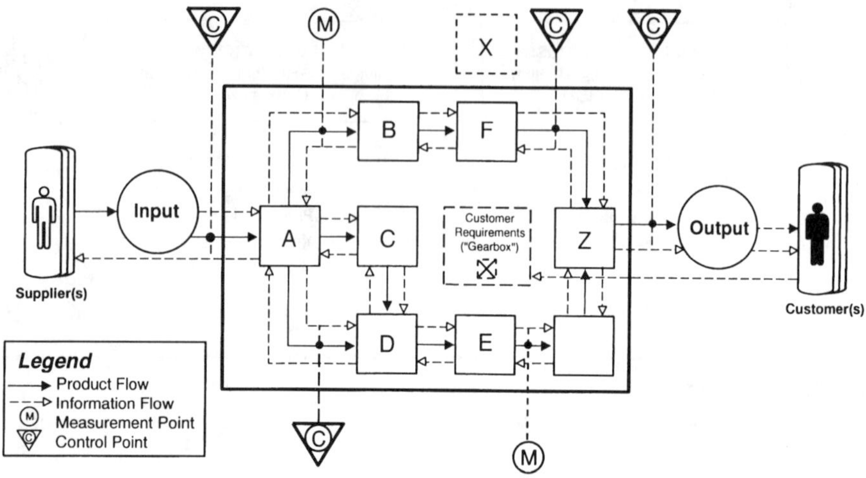

Figure 12.5. Measurement and Control Points in a Business Process

Process measurements are taken at a number of predetermined *measurement points* within and around the business process. If the measurement data show an unacceptable level of variation from pre-set performance targets (e.g., conditions of satisfaction), appropriate corrective actions must be taken at predetermined *control points* to restore process performance to its intended level. These control points usually coincide with measurement points, although some performance parameters may be measured only for monitoring purposes, as they do not require direct corrective action. This fact is behind the practical rule that every control point is a measurement point, but not every measurement point is a control point. Figure 12.5 shows the flow-diagram representation of a business process, with measurement and control points identified. (Process performance measurement and control are described in more detail in Chapter 14.)

It is important to note that the measurement plan[9] for the business process must include performance targets for effectiveness and all the conditions of satisfaction associated with existing commitments for process and subprocess outcomes. In other words, the measurement plan is the prescription for tracking how well process output is meeting agreed-to conditions of satisfaction at every major hand-off point in the process.

Performance targets for efficiency are aimed at the resources and time consumed by the process. They are usually established internally within the process (as, for example, by the process owner) or by enterprise management to ensure the expected value contribution by the business process. Process performance, in terms of efficiency, does not normally concern the customers of the process, unless they are affected by time delays or have to share the costs of the process in some way.

12.1.5 Process Adaptability

Because process capability represents the performance of a stable process with a predictable distribution of output, action upon the process to improve its capability will result in an altered state or, in other words, an improved capacity to produce output. The improvements applied will become permanently embedded in the process.

Process capability changes (as contrasted to process corrections or adjustments within an existing process design and established capability), by definition, imply some process redesign. Process adaptability means the ability to sense changes in customer requirements and then, through rapid redesign or reconfiguration, create the new or improved process capability to meet the requirements and attain the new conditions of satisfaction. Managing process adaptability is based on three key concepts: (1) managing the workflow, (2) managing the current commitments that drive it, and (3) changing existing commitments or establishing new ones (Figure 12.6).

Based on the performance variables that dominate the process, a hierarchy of process measurements is established such that they map the customer-supplier relationships. Output of every subprocess must be specified, variation of output must be measured and identified, and actions must be taken to ensure ability to conform with the conditions of satisfaction at every step in the process. As a result of this effective process control, aimed at maintaining or restoring the state of stability at any given point in time, the process is always capable of predictable performance. In this sense, capability means output produced with predictable uniformity.

As a result of effective continuous improvement, a stable process can always be brought to a state of effectiveness; that is, actually meeting customer requirements or the conditions of satisfaction. However, a process that consistently meets already existing, agreed-to conditions of satisfaction may not be able to adapt to new requirements; therefore, a change of sufficient magnitude in customer requirements warrants a decision to change one or more existing commitments or establish new ones. New or modified commitments mean new outcomes, which may be attained by (1) either improving the existing process capability, or (2) deploying new capabilities.

In all business processes, a transition toward a state of new capability is only possible if information has been imported into the process. This is the concept of *feedback*, and it means that the process can maintain and move itself to higher levels of capability only with the aid of feedback information. The prerequisite for continued process capability change is the availability of adequate and timely feedback information representing the changes occurring in the environment of the process. By definition, the continuing ability for capability improvement and for the deployment of new capability together represent *process adaptability*. The latter is primarily determined by feedback information conveying changed or entirely new requirements.

12.2 Process Centering

Organizations and, in particular, enterprises responsive to change should function as *a system of adaptable business processes*. This is the fundamental concept behind

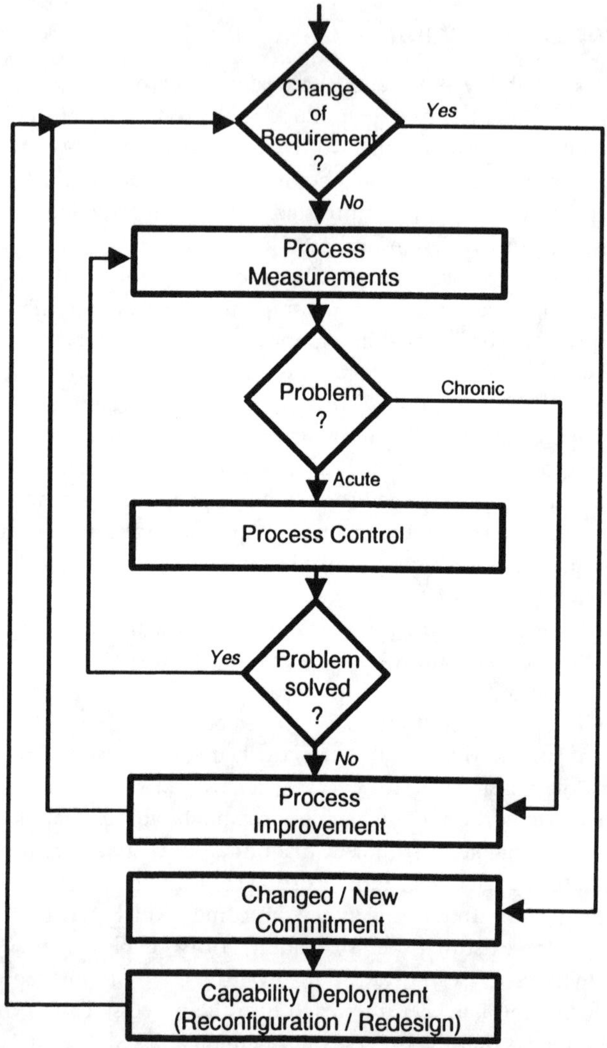

Figure 12.6. Operational Business Process Adaptability

process centering and the alignment of the enterprise around its key business processes. In building the process-centered enterprise, four major stages of managerial tasks have to be completed:

1. *Process thinking, including the conceptualization of the world as a collection of processes forming a virtually limitless network of customer-supplier relationships.* Process thinking views the individual process first through its relationships and interfaces to its external environment — also populated by

processes. Only then does process thinking turn to examining the individual business process, giving way to process management with the prime purpose of understanding and changing the nature, attributes, and workings of the process. Senior management seldom focuses on individual business processes; this is usually left to the business process owner and the stakeholders of the process in question. When senior management is involved in the selection of key business processes for process management (which includes process reengineering), it views individual business processes primarily through their effect on overall business performance.

2. *Redesigning (reengineering) individual processes for optimal current performance and adaptability.* This is the first phase of process management. It involves the rationalization and realignment of workflow and information flow within a given process — the central approach of traditional reengineering. This approach typically involves:
   a. Defining a process by mapping it out as a procedural workflow
   b. Identifying the customers of the process, their needs, and the way those needs are communicated to and captured by the process
   c. Completing the process map by adding the flow of customer-supplier communications (primarily customer requirements and the commitments in response to them) and thus creating the integrated process flow
   d. Establishing various measures and metrics for the measurement of process performance, then, using available measurement data, analyzing and evaluating it
   e. Designing an improved or "to be" process
   f. Planning the implementation of the new design
   g. Operationally implementing and managing the process

   For maximum effectiveness, traditional reengineering of a process is usually carried out by a multidisciplinary, multifunctional process management team, representing all of the functions and disciplines involved in the business process. As noted earlier, success of this approach also requires that the business process team be led by the process owner, who is accountable for the outcome of the reengineering effort and the performance of the newly designed process. The critical role that senior management plays at this stage is (a) the selection of individual processes to be reengineered; (b) the assignment of process ownership, and responsibility and accountability for the outcome of the effort; and (c) providing the necessary resources and authority for successful implementation.
3. *Aligning individual, redesigned business processes for maximum leverage on overall business performance.* Because processes do not operate in isolation and each process affects and is affected by virtually all of the major business processes in the organization, the cumulative effect of reengineering individual processes can only be optimized through purposeful alignment of the individual process goals, commitments made, and outcomes measured. Alignment in this context means that:

a. Process goals are established such that they mutually reinforce and enhance each other.
b. Inconsistencies and contradictions among goals and commitments are eliminated.
c. Process performance measurements are causally linked and harmonized for easier interpretation and analysis.

At this stage, a new order of complexity is added to traditional business process reengineering that involves the interplay of all reengineered processes so that their aggregate leverage on overall business performance can be optimized. Business process alignment is even more important when responding to unexpected change that affects the entire organization. Without the appropriate alignment, the process paradox will come into play: individual process improvement accompanied by overall organizational decline. The corporation now sees itself as a complex system of interconnected business processes that must be redesigned concurrently or at least with concurrent outcomes in mind. While the reengineering of individual processes is a necessary condition, only the coordination and alignment of those processes is sufficient for ensuring that individual process success translates into corporate success. Therefore, enterprises align themselves around the most value-producing work — their *core* business processes — in order to maximize the gain from having rapidly and effectively responded to change. In the alignment stage, senior management's tools include:

a. Coordination of individual process commitments to ensure attainment of the desired enterprise response
b. A system of balanced planning and measurement, such as the balanced business scorecard (BBSC), a powerful tool which can be used for the all-encompassing assessment of corporate performance, as it provides a tangible link between process measurements and corporate performance
c. Enterprise value growth assessment related back to the value contribution of individual business processes

4. *Establishing company-wide learning loops and knowledge base for continued and improved adaptiveness.* Based on the work of the first three stages, this stage links reengineering and process alignment to enterprise knowledge management. The enterprise knowledge base, when properly designed, built, and nurtured, will become the engine of adaptability and response to further, future change.

Once individual business processes are reengineered and the process centering is designed, the last phase of business process management is operational implementation. We will see in Chapter 18 that the two key objectives at the process level are to ensure process effectiveness and efficiency in current terms *and* maximum adaptability for dealing with future eventualities.

12.3 Summary

The two most important objectives of operational process management are to ensure (1) process effectiveness and efficiency in current terms, and (2) maximum adaptability for handling future change. To achieve these, effective process management must focus on the interactions of the process parts: the series of commitments and communications that drive the actions of those parts. The key to adaptability is how commitments are made and managed, how they drive the work done, and how much customer value that work adds. The two dimensions of operational process management, then, are (1) managing commitments, and (2) managing actions (work activities).

All the enterprise's work is done in its business processes. When companies reengineer, they redefine the flow of communications and work that the people in the various organizational units and departments do. It is processes and not organizational units or departments that are the object of reengineering.

The operating process consumes quantifiable amounts of resources in producing its output. These resources ultimately fall into five asset categories: people, materials, energy, equipment, and information and knowledge (including methods). Establishing and using the correct measures are crucial to the success of managing business process performance. Appropriate measurements provide the basis for determining how well customer requirements are met (effectiveness) and how much resource and time are consumed in doing so (efficiency). Managing process adaptability is based on three key concepts: (1) managing the workflow, (2) managing the current commitments that drive it, and (3) changing existing commitments or establishing new ones.

Once the adaptability of key business processes is assured, the enterprise can align itself around the most value producing work, its core business processes. Alignment is accomplished through two major stages of managerial action: (1) aligning individual processes for maximum leverage on overall business performance, and (2) constructing an enterprise-wide network of learning loops and knowledge base as the intellectual prerequisite for continuing adaptiveness. Once all this planning work is done, the enterprise is ready for operational implementation.

13 Process Structure: The Holistic View

In the previous chapters, we have established that effective process management must be based on coordinating a series of commitments that drive process work activities. We have also concluded that, as a consequence, process management has two dimensions: managing commitments and managing work activities. We will now apply a holistic model to examine what structural arrangements are needed within the boundaries of the process to make this happen.

The holistic process model views business processes as structures where information organizes the flow of work and its outcome. The construct of the customer-supplier dialogue provides the structural base for the disciplined negotiation, communication, and management of commitments. The resulting network of commitments and its inherent reconfigurability are key design features for process adaptability and indispensable management tools for ongoing process adaptation during the operational life of the process.

From a methodological point of view, process centering comprises (1) individual process reengineering to achieve adaptability, and (2) alignment engineering. The holistic approach overcomes three well-known limitations of traditional workflow-based reengineering:

1. The mechanistic view defines a business process as a repeatable sequence of work activities performed by a defined set of resources over time. With this mechanistic focus, traditional reengineering assumes an inherent preference for realignment and restructuring with the purpose of optimizing the process performance around key parameters such as quality, cost or cycle time — one business process at a time.
2. Because of its selective focus on the individual business process, traditional reengineering provides no design mechanism for integrating the new process design with the corporation's strategic business goals or strategies.[1] This

limitation, in combination with the first, may lead directly to the process paradox because, while achieving the lowest cost or the shortest cycle time for a given process is a notable achievement, it may or may not measurably help improve overall enterprise performance.

3. The sources of innovation and growth seldom, if ever, lie in repetitious, mechanical arrangements of chronological activities; rather, they are uncovered in dynamic and responsive customer-supplier relationships across major business processes within the enterprise. These relationships can provide the critical cause-effect mechanisms needed to deal with *ad hoc* situations, where every new request received, every new offer made can represent an idea generating a new solution.

13.1 The Dynamic Business Process

What is needed, then, is to move from viewing and managing business processes as mechanical work structures that generate information, to modeling them — from design through implementation — as information structures capable of organizing the flow of work.[2] In this world view, information organizes work and matter into a formal structure, resulting in physical work and material flow.

The prime function of information in this model is that of communication and direction through commitment. Information also determines and describes the flow of work and deployment of resources. When the information changes, signaling change in the environment or within the process, a new capability and a new workflow and resource structure will evolve. The business process becomes an information network that integrates work, matter, and information itself based on a continuing series of agreed-upon commitments. It keeps evolving and changing as it captures, processes, and accumulates more information.[3] In a dynamic business process, information manages work and its outcome. For a process to remain adaptable and competitive, information must be continually generated, renewed, and accumulated into a knowledge base that, in turn, becomes the intellectual capital of the enterprise. The chain of customer-supplier relationships within the process is, of course, the mechanism through which empowered and accountable people generate and manage the information needed for managing an effective, efficient, and competitive business process.

13.1.1 Process Definition

In line with Jantsch's concept, an integrated process flow does not mean the mere augmentation of procedural activities by adding communications; what it does mean is that a set of negotiated and agreed-upon commitments drives a series of capabilities (procedures and resources) to perform work and produce the committed outcome. The difference between the two approaches is fundamental: When communications are simply added to the procedural workflow, the result will be a flow of descriptive information always and only related to the process output. The integrated process flow, on the other hand, includes the entire set of customer-supplier communications, including all the give-and-take of negotiations between the two parties.

The concept of integrated process flow, in itself, is not new: a number of authors and practitioners have described several versions of it under various names beginning as early as 1987.[4] What is new, however, is the recognition of its unique contribution to process adaptability and its natural implementability in a networked information environment.

Most process design methods focus only on activities and largely ignore the communications driving them. Traditional process definition revolves around the question of output: "What is produced by what activity using what resources and in what sequence?" The holistic definition, on the other hand, provides answers to two questions. First, it answers the question of commitments: "Who has committed what outcome to whom?" Second, it addresses the question of outcomes: "What capabilities produce the committed outcome to meet what conditions of satisfaction?" The fundamental difference here is that the traditional definition starts with *what is*, the holistic approach with *what needs to be done.*

In the holistic approach, the process definition is different from the traditional method used with procedural workflows in that the governing customer-supplier relationships must be established first. The desired outcome of each relationship then leads to the definition of the procedural activities required to produce it.

The traditional process definition assumes the process to be static. It then proceeds, by using decomposition, from the macro level to successively more detailed levels of the workflow. The holistic definition, on the other hand, considers the process to be dynamic. It starts at the business process (macro) level by defining its governing customer-supplier relationships with their resulting network of commitments, associated outputs (deliverables), and related accountabilities. By creating accountabilities associated with each commitment in the customer-supplier relationship, the holistic approach naturally integrates humans into the process flow and thereby provides one of the key management components of process adaptability.

From this point, the procedural workflow is composed from the sets of capabilities required to produce each output — usually at the subprocess level. There is no work, as such, flowing between subprocesses, as they are linked by means of (1) coordinated commitments, and (2) the flow of interim outputs among them, until the final output is delivered by the overall process. In summary, holistic process definition starts with customer-supplier relationships at the overall process level and with procedural workflow at the subprocess level.

In the following, we will describe a rigorous approach to dealing with commitments made and managed by empowered and accountable people. First, a holistic process model[5] for a single customer-supplier relationship is introduced. As we will see in Chapter 14, this model then becomes the basis for (1) the holistic design of the business process, and (2) defining and distributing accountability in the process. Both are key elements of the design for adaptability.

Phase	Customer	Supplier
	Communication Acts	
DEFINITION	Request (Need)	Offer
NEGOTIATION (of conditions of satisfaction or changed requirement)	Counter-request or Commit (Withdraw)	Counter-offer or Commit (Withdraw)
PERFORMANCE (against requirement)		Report completion (Withdraw)
ASSESSMENT	Accept or Reject (Withdraw)	

Figure 13.1. The Customer-Supplier Dialogue (Adapted from Scherr, A.L., A new approach to business processes, *IBM Systems Journal*, 32(1), 83, 1993. With permission.)

13.1.2 The Holistic Process Model

The holistic approach defines the business process in the framework of a dialogue that takes place whenever two parties conduct business. The dialogue occurs over time, and it comprises four chronological phases which are representative of any business transaction completed between a customer and a supplier.[6] Each phase contains one or more communication acts.[7] (The customer-supplier dialogue is illustrated in Figure 13.1). The four phases and the communication acts occurring in each, are

1. *Definition phase:* The customer defines a desired outcome and then communicates a *request* to the supplier, or the supplier defines a potential outcome and then communicates it as an *offer* to the customer.
2. *Negotiation phase.* Various conditions of the delivery of the outcome (regardless of whether it was initially requested or offered), such as quality, price, timing, payment by the customer, and so on, are negotiated and refined through one or more iterations. During these iterations, the supplier can counter a request with a *counter-offer*, and a customer can counter an offer with another request until agreement is reached. The conditions agreed to by both parties are the *conditions of satisfaction* to which both parties commit themselves. This mutual *commitment* then becomes the driver of all that happens in the remaining two phases.
3. *Performance phase (or work phase)*: The supplier fulfills the conditions of satisfaction by performing the work required to produce and deliver the

products or services that were committed at the end of the negotiation phase. The supplier then *reports* completion to the customer.

4. *Assessment phase:* The customer assesses the supplier's work product and declares satisfaction by *accepting* it. Along with this acceptance, the customer's part of the commitment (e.g., payment of invoice) is also fulfilled.

The communication acts included in the above description are *request* or *offer, commit, report,* and *accept.* They occur when the four phases progress smoothly; however, depending on the prevailing conditions, either party can *withdraw* from the dialogue and thereby terminate it at any point in time. Also, the customer can *reject* the supplier's work product in the assessment phase, if it does not meet the agreed-upon conditions of satisfaction. In this case, the customer may make a new request and start another dialogue, or withdraw. Similarly, the supplier whose work product was rejected may make a new offer to re-work or replace the work product, to change the conditions of satisfaction, or withdraw. The set of all possible communication acts and the rules governing their use are known as the customer-supplier communication *protocol.*

A single execution of the four-phase dialogue represents one completion of the business process and is called a *transaction* or *business cycle.* A transaction is either terminated or it must be carried to completion — there are no open transactions.[8] The transaction is terminated when one of the parties has withdrawn. A transaction is complete when there are no unfulfilled commitments left — that is, all of the commitments made during the transaction have been satisfied. There are no "surviving" commitments between transactions.

This makes the customer-supplier dialogue *complete*; it represents all possible outcomes and events of a business relationship. From the customer's standpoint, the customer-supplier dialogue is also *satisfying*; unless one of the parties withdraws, a transaction does not end until the customer declares the work to be satisfactorily completed. Finally, the dialogue is *scaleable*; it can describe equally well enterprise-to-enterprise relationships or simple transactions between two individuals.

This four-phase dialogue applies not only to conversations with external customers of the enterprise, but also to all related internal customer-supplier relationships which, when integrated with the work flow they drive, will form the whole business process. In such a process construction, the customer-supplier dialogue becomes the standard interface between related work capabilities usually identified at the subprocess level. The two key results of the dialogue, the commitment (C) and the output (O), are highlighted in the customer-supplier interface shown in Figure 13.2.

In the holistic model, each of the four phases of the customer-supplier dialogue is considered a unique series of activities and events associated with definable, but potentially different, sets of capabilities. The sum total of these capabilities at any given time represents the outcome-producing capacity of that particular customer-supplier relationship.

Beyond these, each phase can generate secondary dialogues that define interactions with other participants in the process. Each such dialogue has its own four-phase event structure with related capabilities and its own conditions of satisfaction.

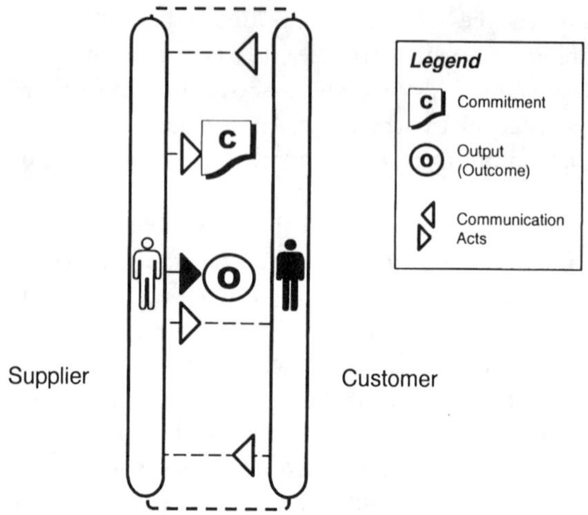

Figure 13.2. The Customer-Supplier Interface

Secondary dialogues can be decomposed into further subordinate ones until the roles of all participants in the process have been defined in a complete network of dialogues. However, as subordinate dialogues are defined, they must be linked to an already defined dialogue by a common party playing the alternative roles of customer or supplier. For example, when a supplier initiates a secondary dialogue to request information or a secondary commitment for responding to a customer request, he or she becomes the customer in the secondary dialogue. (Figure 11.2.c illustrates this concept; the human icons in each of the subprocesses represent the same individual, but in two different roles.)

Using the holistic business process model, a simplified version of the order-fulfillment process of Union Camp's Fine Paper Division could be described as follows:

1. The customer enters an order (request) for a shipment of fine paper with the Union Camp customer service representative. Initial conditions of satisfaction are the quantity, price, and required date of shipment. This establishes the primary customer-supplier relationship between the customer and the representative who acts on behalf of the Fine Paper Division.
2. The representative checks the requested price against the published price list. If this matches the customer's request, a delivery commitment is requested from the inventory function. (If there is a price mismatch, the representative will attempt to negotiate a new price with the customer or with the supplier which will modify the initial conditions of satisfaction.) This establishes a secondary customer-supplier relationship between the customer service representative and the inventory function.

3. The inventory manager checks both the current inventory level and ongoing paper production in the mill (thereby establishing a tertiary relationship with the mill) to determine from which source (or from what combination of sources) the requested quantity of paper can be supplied by the required time. Based on the actual situation, the inventory manager may agree to the representative's request and commit the required quantity, or make a counter-offer.
4. With the inventory manager's commitment in hand, the customer service representative requests shipment through the transportation function by the required date. This establishes another secondary customer-supplier relationship. Based on the availability of transportation capacity and after consultation with several transportation companies (another tertiary relationship), the transportation manager may agree to the representative's request and commit the delivery date, or make a counter-offer.
5. The customer service representative conveys the commitment (or counter-offer) to the customer and eventually agreement is reached which then represents the agreed-to commitment to fulfilling the customer's order.
6. The report of completion from the transportation company to the representative concludes that particular secondary transaction.
7. Upon receipt of the transportation report, the representative reports order fulfillment to the customer and presents an invoice.
8. The customer assesses the shipment received, conveys acceptance, and pays the invoice.

In this example, the customer-supplier chain includes the following parties:

1. Customer
2. Customer service representative
3. Inventory function
4. Paper mill
5. Transportation function
6. Transportation companies

The primary customer-supplier relationship between the customer service representative and the client drives all others. In the process described above, there are two secondary customer-supplier relationships: those between the customer service representative and both the inventory function and transportation function. There are also two additional relationships of importance: the inventory function/paper mill and the transportation function/transportation company.

As we have said, every subordinate dialogue has its own four-phase structure, unique conditions of satisfaction, and self-contained capability. Because of this, each subprocess associated with defined parameters can be considered as an autonomous management unit having an empowered owner accountable for the result it produces.

The holistic process definition includes communications as well as work activities. Accordingly, it is not a replacement of the traditional input-process-output

Table 13.1. Comparison of Process Model Attributes

Attribute	*Input-Process-Output (IPO) Model*	*Holistic Model*
Business environment	Stable	Changing
Purpose	Produce quality products efficiently	Fulfill commitments
Orientation	Output	Customer
Boundaries	Static	Dynamic
Structure	Activities	Dialog
Connecting mechanism	Inputs/outputs	Customer-supplier relationships
Nature of work	Processing	Conducting business
Key measurements of satisfaction	Effectiveness and efficiency	Fulfillment of conditions
Approach to quality	Process control and improvement	Accountability
Role of Information Technology	Perform procedures	Coordinate commitments and actions

Source: Adapted from Center, K. and Henry, S., *A New Paradigm for Business Processes,* IBM Corporation, Austin, TX, 1992. With permission.

(IPO) model (Figure 11.2.b); instead, it incorporates it. The IPO model continues to be used for describing procedural activities occurring within each of the four phases of the holistic model. Thus, both the customer-supplier communication protocol and the IPO model are needed for a complete, holistic description of the business process. A comparative summary of the various attributes of the IPO model and the holistic model is shown in Table 13.1.

Defining business processes as more than procedural workflow can help management focus on processes that may be in dire need of attention. When processes are viewed as information-driven series of activities, then such processes as governance, acquisitions and divestitures, litigation, management succession, business planning, research, and other creative processes can also become the object of effective process management. Such processes may lack obvious inputs such as raw materials, parts, documented orders, and the like — or tangible outputs such as goods produced, reports released, etc. Recognizable workflow patterns may be missing; yet, they can and should be subject to planning, control, improvement, and adaptation. Such processes are recurrent but not necessarily repeatable in every detail of their sequence. They may have major impact on organizational performance; in each instance, they may be performed in different ways that make a difference in terms of the resources consumed. But such processes always have one common attribute: an extremely heavy content of customer-supplier interactions and commitments.

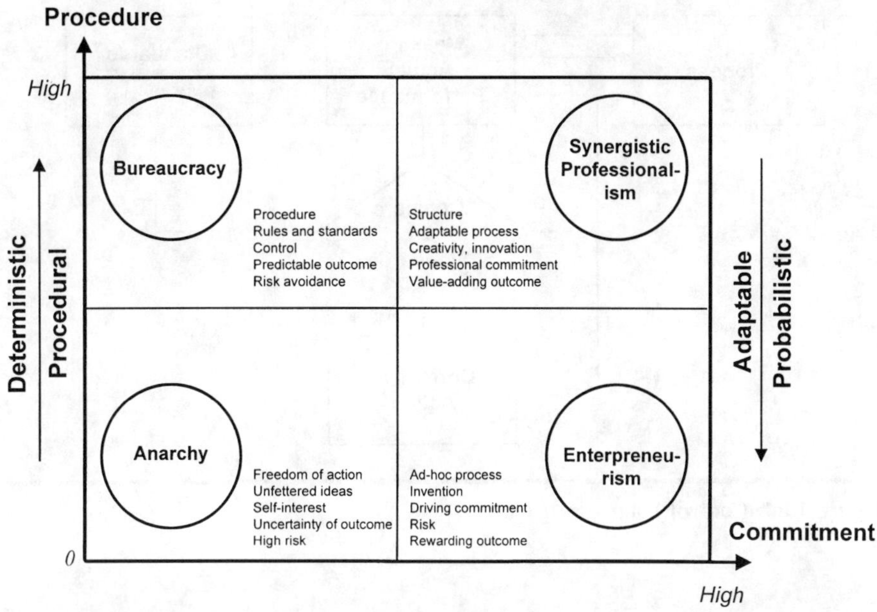

Figure 13.3. The Domain of Business Processes

In terms of event and work content, business processes range between two extremes: (1) completely predictable, repeatable procedures with a pre-determined workflow and no defined commitments and associated accountabilities; and (2) entirely probabilistic, often *ad hoc* processes depending on well-defined but changeable commitments and accountabilities. Clearly, the holistic model covers the entire range with most business processes in real life, falling somewhere in between and exhibiting a certain mixture of commitment and procedural content (Figure 13.3).

Examples of entrepreneurial processes with well-defined commitments and virtually no procedures include marriage and parenting, the startup of a new business, research, nursing, coaching and counseling, and the creation of objects of art. Processes with well-defined procedures but no commitments are usually just highly developed bureaucracies that are mostly rule-driven procedures, such as applying for a Social Security number, filing a tax return, renewing a driver's license, discrete manufacturing, product testing, etc. Examples of synergistic, professional processes with both well-developed commitments and procedures are an emergency room team, a football team, a symphony orchestra, and an airliner cockpit crew, among others. The ideal process, however, may not be one with both commitments and procedures fully developed; for most enterprises, the optimum is somewhere within the spectrum — depending on purpose and the ability to generate maximum value.

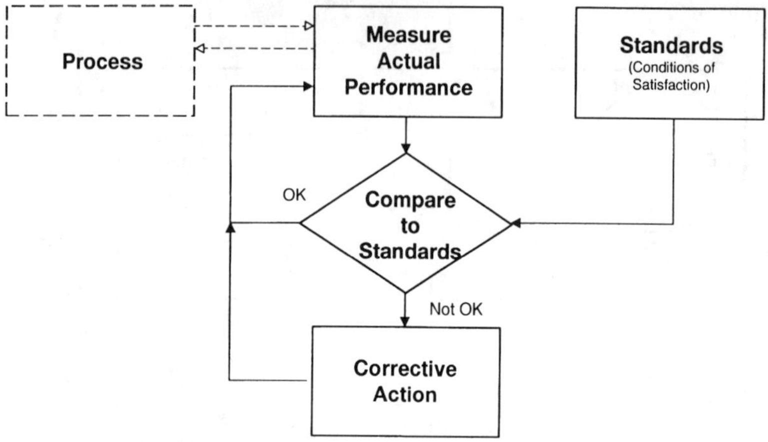

Figure 13.4. Control Loop

13.1.3 The Workflow and Adaptive Loops

The workflow associated with each phase of the customer-supplier interaction includes all the activities which lead to the communication act that completes the phase. The performance phase also includes those procedural activities required to produce and deliver the work product to the committed conditions of satisfaction. In each phase, the work is usually a mix of highly deterministic, procedural sequences and more information-intensive, knowledge-based activities. It may also include one or several adaptive loops[9] discussed earlier. They are the drivers of purposeful actions that change physical reality.

A special case of the adaptive loop is known as the *control loop* (Figure 13.4), where the decision is binary (act or not to act) based on evaluation of the information sensed. Control loops are routinely inserted into procedural workflow to ensure its stability and effectiveness while producing outcomes. This is the well-known management mechanism of process control.

Another version of the adaptive loop is the *learning loop* (Figure 13.5), where a relatively large number of decisions can be made, based on the interpretation of the information gathered. Learning loops are used to capture the results of knowledge work and to guide its further progression; they are also applied to monitor the flow of information between customer and supplier and to capture the content and chronology of the customer-supplier dialogue. Learning loops change mental states but do not directly change the physical world.

From a structural standpoint, an adaptive process is an open system containing learning loops. Because self-organization arises from intelligence and learning, it can reach higher states of organization as a result of learning derived from the information fed back into the system from within or from the external environment. Adaptive

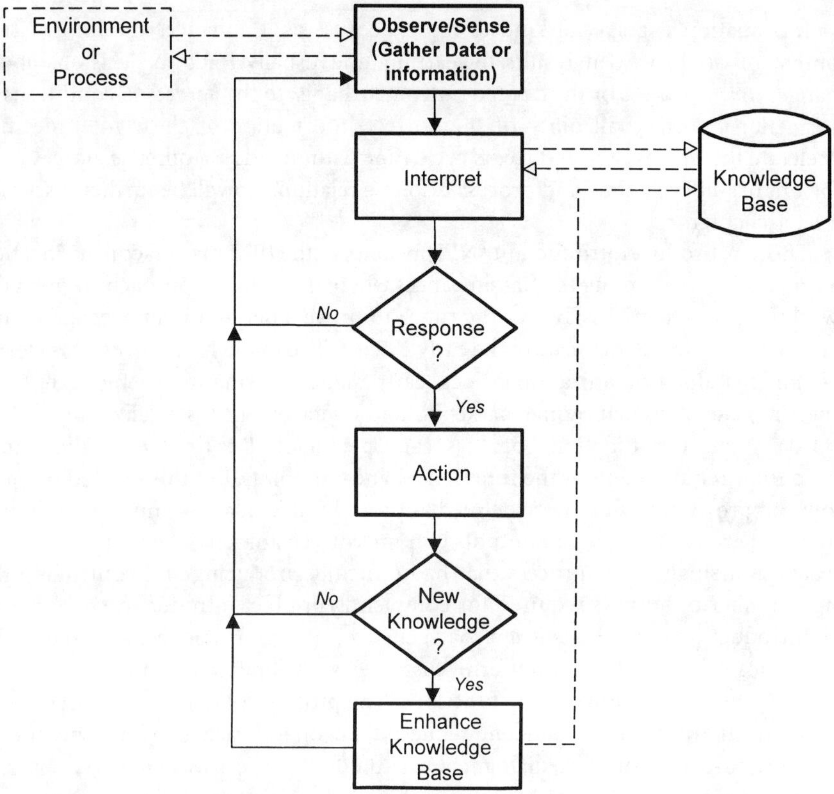

Figure 13.5. Learning Loop

processes are a special class of open systems; they are open to incoming information, but closed to the exchange of matter or energy until a decision within the system is made to undertake such an exchange. In a business process, such a decision is always made as a commitment to achieve customer satisfaction. A simple example of such an adaptive process is the room service in Pan Pacific Hotels, where the service manager is empowered to obtain items from outside suppliers if a customer orders food or beverages not on the hotel's menu or not readily available from the hotel kitchen.

Inputs to the process can include increments of any and all of the process resources: people, materials, energy, equipment, or information. All of these inputs come from the environment external to the process; some, such as materials, energy, or equipment, come from external suppliers. In any case, process inputs can be thought of as resource enhancements to the capabilities associated with the procedural workflow. The unavailability of the correct resources when required will result in delays. Process cycle time, the time it takes to perform the work activities required to produce the desired outcome, is dependent on having the proper resources in place at the right time.

Information is a special kind of input because as it enhances the information content of the process it results in learning and usually leads to decisions about changes to the process or the desired outcome (that is, to the agreed-to conditions of satisfaction). Also, availability of the correct information at the right time may accelerate the process and reduce its cycle time without adding other resources to it. The important cycle time and process resource relationship will be further discussed in later chapters.

The new product introduction (NPI) process at the IBM manufacturing location in Austin, TX, was among the first instances where the holistic approach (termed the "workflow paradigm" by IBM at the time) to process definition and reengineering was used on an industrial scale in the early 1990s.[10] The initial NPI process was clearly fractionated along organizational lines: early manufacturing involvement, test engineering, assembly engineering, marketing, and several others. Essentially using an IPO approach, the cross-functional process team spent a lot of its time reconciling input and output requirements at the functional hand-off points, but this was tedious and slow. Subprocess activities were defined as procedural with a large number of inputs from numerous suppliers and outputs to numerous customers. The NPI process was treated as a single mega-process that had difficulty producing and controlling the large number of outputs required for completing product introduction.

Introduction of the holistic approach changed all this. It was recognized that the NPI process is actually a long-term customer-supplier relationship between the development and manufacturing functions. The process contains a very large number of related and interdependent requests and offers, each of which initiates a complete customer-supplier dialogue. Once the NPI process was redefined in these terms, the process team was able to identify specific dialogues that drove and coordinated the various activities of the participating functions and organizations. Individual employees came to the realization that they were players in several distinct and definable customer-supplier relationships, not cogs in an enormous process that produced just a large array of outputs.

Another example of redefining the NPI process is what Gennum Corporation (Toronto, Canada) did to deal with product launch challenges. When Gennum, a manufacturer of microchips and microprocessors for the communications and hearing instrument industry, embarked on the reengineering of its NPI process in 1995, it had an experience similar to that of IBM in Austin. As the process management team was using the traditional IPO approach to process definition and redesign, they were frustrated by the IPO model's limitations in dealing with the high level of give and take, the rich flow of requests and offers between the major functions.

Eventually, the Gennum team abandoned treating the NPI process as a single, massive business process. They simply could not effectively deal with the unfathomable variety of inputs and intractable assortment of possible outputs, from ideas and draft designs through test results to the marketable physical product. Instead, they broke up the process into three major subprocesses, each of which was defined as a mini-enterprise tied to the others in the form of true customer-supplier relationships. The mechanism for these relationships was a form of dialogue (known as "gating"), an interface created, in essence, for decision-making and commitment tracking.

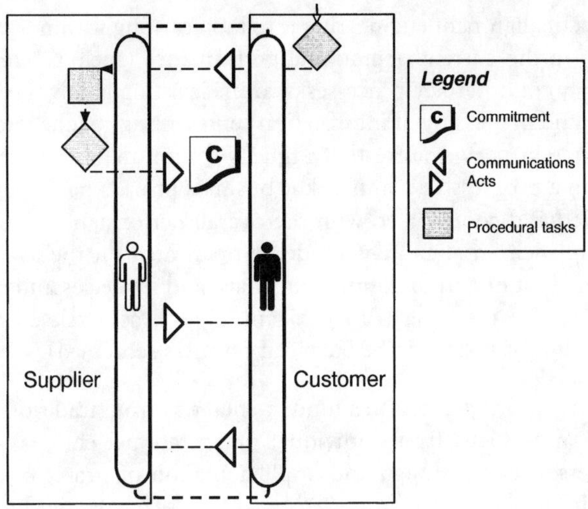

Figure 13.6. Communication of Commitments

13.2 Alignment Engineering

The number of existing or potential adaptive loops within an organization can be extremely large; therefore, defining all of them would be an impractical and virtually hopeless exercise unless one is using custom systems software, which requires a lot of planning and preparation but eventually gets the job done. Without the software, it is more practical to first define adaptive loops across the key business processes that add maximum value to the corporation's overall performance. In essence, these adaptive loops link strategically related commitments across the core processes. When a change is sensed in one commitment (delay, inability to deliver, and so on), the interpretation of this information may lead to a decision to renegotiate it or to establish a new dialog for achieving an entirely new outcome. This approach, nevertheless, generates two new dimensions to process redesign:

1. In the reengineering of individual business processes, the redesign focus moves from the workflow to the information flow within the process, and especially to the communication and management of commitments between each customer/supplier pair in the value chain. The design result changes from a chronological sequence of events to a dialog that is based on cause and effect. This will create the elemental base for tracking the status of interrelated commitments within and across business processes (Figure 13.6). This is key to achieving process adaptability.
2. Emphasis shifts from individual processes to the mutual alignment of multiple processes and the inter-relationships among them. Designing and implementing such an alignment are known as alignment engineering. A key task

> key task of alignment engineering is to link learning within one key business process to the current operational performance (such as effectiveness and efficiency) of other key processes. Process learning involves observing, capturing, and processing of information representing change in commitments or change in performance that impacts commitments within the process. Then, in a progressive manner, key business process performance becomes identified and coordinated with the overall performance of the enterprise. Building such an alignment model is predicated on the availability of an integrated set of corporate mission, goals, and strategies and an enterprise-level resource planning (ERP) system, as well as a related measurement system (as, for instance, the balanced business scorecard).

Alignment engineering differs in a fundamental way from traditional reorganization in that it shifts focus from individual organizational entities (departments, divisions, processes) to the design and implementation of strategic adaptive loops among key business processes. Figure 13.7 shows the use of adaptive loops to link the three subprocesses A, B, and C within a larger business process. Loop B is established to capture output status data in subprocess B, because the output of this subprocess is needed for making the commitment in subprocess A. At the same time, the commitment in subprocess C is tracked through loop A as it, too, is a prerequisite to making the prime commitment in subprocess A.

13.3 Summary

The holistic process model views business processes as information structures capable of organizing the flow of work and its outcome. Because of this, the holistic approach can overcome several well-known limitations of traditional workflow-based reengineering. The construct of the customer-supplier dialogue provides the structural base for the disciplined negotiation, communication, and management of commitments. The resulting network of commitments and its inherent reconfigurability are key design components for process adaptability and indispensable management tools for ongoing process adaptation during the operational life of the process.

Most process design methods focus only on activities and largely ignore the communications driving them. The fundamental point here is that the traditional definition starts with what is, the holistic approach with what needs to be done. The holistic definition considers the process to be dynamic. By creating accountabilities associated with each commitment in the customer-supplier relationship, the holistic approach naturally integrates humans into the process flow and thereby provides one of the key management components of process adaptability. It defines the business process in the framework of a dialogue which occurs over time and it comprises four chronological phases. Each phase contains one or more communication acts between the customer and the supplier.

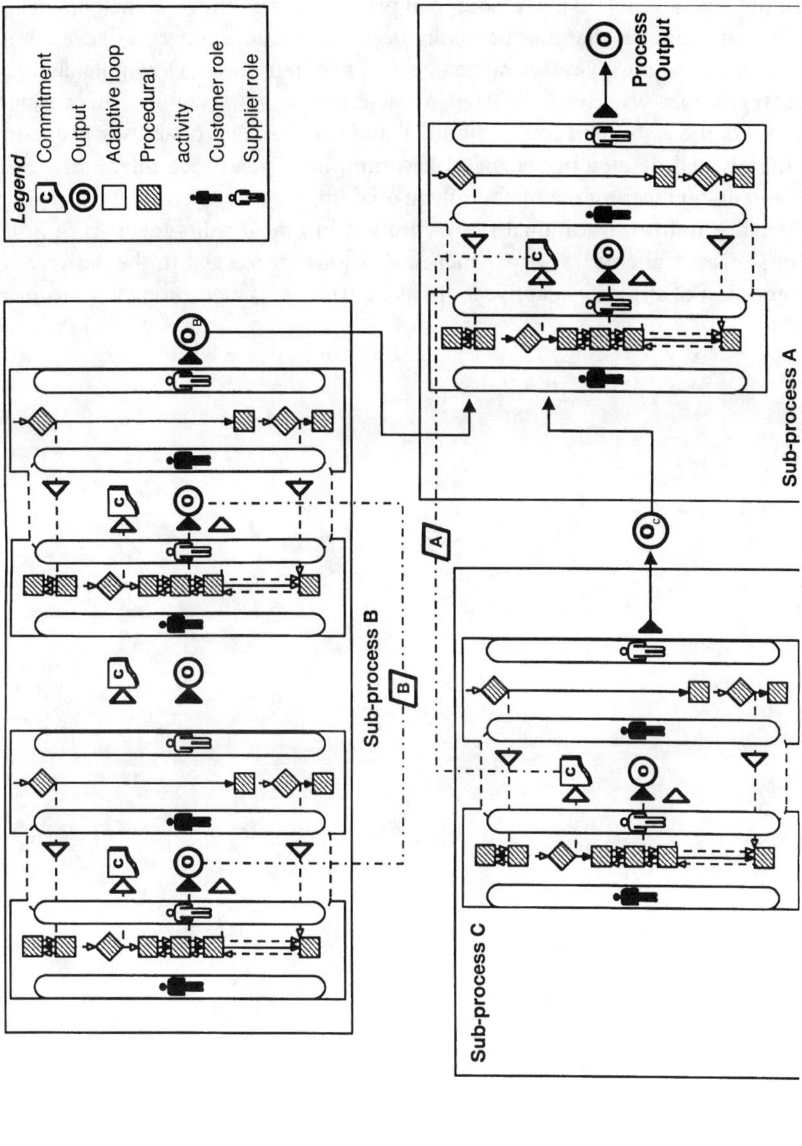

Figure 13.7. Use of Adaptive Loops for Linking Commitments and Outcomes

A single execution of the four-phase dialogue represents one completion of the business process; it is called a transaction or business cycle. The dialogue applies not only to conversations with external customers, but also to all related internal customer-supplier relationships which, when integrated with the work flow they drive, will form the entire business process.

Defining business processes as more than procedural workflow can help management focus on processes that may be in dire need of attention. In terms of event and work content, business processes range between two extremes: (1) completely predictable, repeatable procedures with a pre-determined workflow and no defined commitments and associated accountabilities, and (2) entirely probabilistic processes depending on well-defined but changeable commitments and accountabilities.

Alignment engineering is a methodology used in process centering. It differs in a fundamental way from traditional reorganization in that it shifts focus from individual organizational entities (departments, divisions, processes) to the design and implementation of strategic adaptive loops as information links among key business processes.

14 Process Performance and Resources

All work is done in processes; therefore, performance measurements should be applied to the products of the process and to process work itself. The design of performance measurements covers both dimensions of managing commitments and managing the procedural workflow. Cycle time reduction, particularly with respect to the time to respond, is another key component of the design for process adaptability.

The holistic approach to process definition and management requires an empowered human resource organization, accountable ownership, and uniquely tailored capabilities in the information technology component of the process. Information technology is designed to accelerate both communications and the flow of work and information in the process and to integrate, within the process and across the enterprise, the management of commitments and outcomes.

In most traditional organizations, the business process work is done in a chopped-up manner. The performance of the individual divisions, departments, and sections performing their particular piece of the process is usually measured in a number of ways, but always against their functional or discipline-oriented objectives. Most functionally organized companies are barely aware of their business processes and certainly possess no quantifiable understanding of their performance.

Nevertheless, the work exists and is being done; processes also exist, but they are just "rolling along" without any management attention paid to them. Because it is the process where work is done, it is logical to conclude that performance measurements are most meaningful when applied to the products of the process and to work itself.

14.1 Performance Measurement and Control

What gets measured, gets done; what is done, can be changed or improved. This adage sums up the essence of process measurement, control, and improvement. Measurements and controls are designed to ensure that management can always determine the level of performance in the process and minimize or eliminate unwanted variation in that performance. Adaptability, on the other hand, means the ability to introduce desirable changes into the process rapidly when required. This can always be done to a process with stable and known current performance.

14.1.1 Process Measurement

The input-process-output (IPO) model uses traditional measurements of process performance: effectiveness, efficiency, cycle time, and process cost. In the holistic process model, these will still apply to the procedural workflow and the output it produces; however, there is now a second set of performance measurements to consider — those related to managing and communicating commitments in each customer-supplier relationship within the business process.

If the customer-supplier dialogue is flawed; if communications lack accuracy, truthfulness, or intellectual honesty; if it takes too long to arrive at an agreement; or if there are too many withdrawals during any of the dialogue phases, this may have just as much or even more impact on the outcome as defects in the procedural work. Deficiencies in the management of commitments are detected by means of learning loops — control loops applied to the information flow within the process (Figure 14.1). Process improvement or redesign opportunities are then identified by means of analyzing communication patterns that lead to either highly satisfactory or very poor results.

The holistic model structurally incorporates the measurement of customer satisfaction in its assessment phase at the completion of each business cycle. This instantaneous, real-time availability of customer feedback can provide considerable flexibility to process managers in modifying the procedural workflow in time for the next cycle, a significant competitive advantage to the enterprise.

The consumption of process resources is the basic indicator of process efficiency across both components (commitments and workflow) and all four chronological phases of the customer-supplier dialogue. The consumption of human time and energy is of particular significance in the definition, negotiation, and assessment phases (see Chapter 13). As we have seen earlier, activities in these phases primarily comprise knowledge work and communications, frequently referred to as "overhead" — indirect or staff work. They consume a significant amount of human resources and produce a large proportion of waste in most corporations.

Time-based measurements include elapsed time for each phase of each customer-supplier relationship (overall dialogue time or transaction time, cycle time for the performance phase, and so on) and the timeliness of communication acts (meeting schedules and deadlines) in each phase.

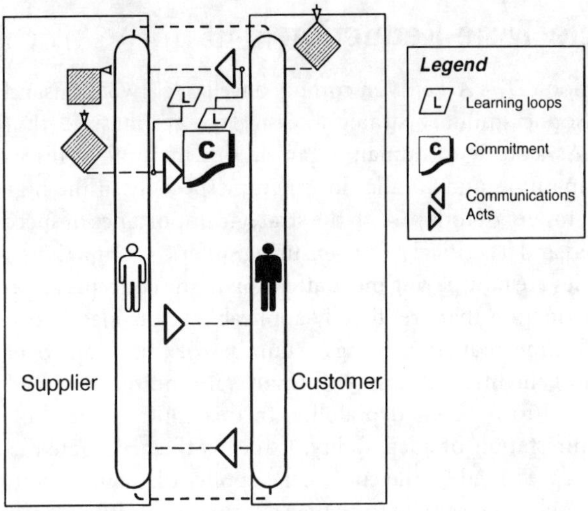

Figure 14.1. Communication of Commitments

14.1.2 Process Control

Process control in the traditional IPO process model is based on a system of strategically located control loops (control points) within the business process. The control loop provides for corrective action after the fact; that is, after a defect in execution (variance from the standard) has been detected. In the holistic model, and especially related to the management and communication of commitments, controls have to be more anticipatory. They usually take the form of empowerment, guidelines, checklists, restrictions on undesirable outcomes, and the like. Honda, for example, wants a process continuously improved so that the variance from standard never occurs.

Quality controls are still applied to the procedural flow to ensure customer satisfaction. However, they play a much more important role as time controls in the form of alerts on overdue procedural actions, communication acts, or prompts for pending actions or escalation (transfer to another customer-supplier relationship or business process) for decision and commitment. Another important category of quality controls ensures the completeness of content, the clarity, accuracy, and precision of the communication acts within the customer-supplier dialogue.

Last but not least, we should point out that the communication protocol of the dialogue is designed to eliminate ambiguity. However, the protocol itself cannot assure the authenticity of the communications. (For purposes of this discussion, "authentic" means that both communicating parties mean what they say, say what they mean, and believe that they know what they mean.) Achieving authenticity is not just a control issue, but a matter of culture and organizational behavior. (This is discussed in more detail in Chapter 16.)

14.2 Cycle Time Reduction

In his famous book, *The Art of War*, compiled well over two thousand years ago,[1] the Chinese philosopher-military strategist, Sun Tzu, emphasized the importance of speed in war. As a study of human organizations in conflict, his work very much applies to competition and conflict in general, especially in the highly competitive business environment of today. With the strategic importance of speed being increasingly recognized and its critical role in ensuring superior competitive position understood, companies are now paying more attention to the concepts and methods related to cycle time reduction that are directly applicable to business processes.

While cycle time reduction brings a number of important benefits to process management and enhances process performance in a number of ways, our focus here is its contribution to process adaptability. In this context, speed of response is an important manifestation of adaptability; it means the accelerated execution of the business cycle represented by the customer-supplier dialogue operating in real time with minimum delay. The ability to change existing commitments or make new ones is of particular significance.

14.2.1 Cycle Time Basics

The two dimensions of the holistic process model, communications about commitments and procedural activities, show different cycle time characteristics and, therefore, call for different design approaches to reduce cycle time.

14.2.2 Business Cycle Time

In a holistic process environment, business cycle time (also called transaction time) is the overall time-based performance indicator. It is defined as the time elapsed between the receipt of the customer's original request and the customer's acceptance of the outcome (Figure 14.2.a). Two components of the business cycle time are particularly important:

1. *Time to commitment:* Time elapsed between the receipt of the customer's original request and the communication act of commitment (Figure 14.2.b)
2. *Performance cycle time:* Time elapsed between the communication acts of commitment and report of completion (Figure 14.2.c)

Time to commitment is primarily a function of the speed and time efficiency of customer-supplier communications during the negotiation phase. Performance time, on the other hand, is almost exclusively determined by the time-related characteristics of the procedural workflow, of which there are several.

14.2.3 Time To Respond

If a business process has to respond to new or changed requirements and/or adapt to unexpected change, the time to respond (also known as adaptive cycle time) is the

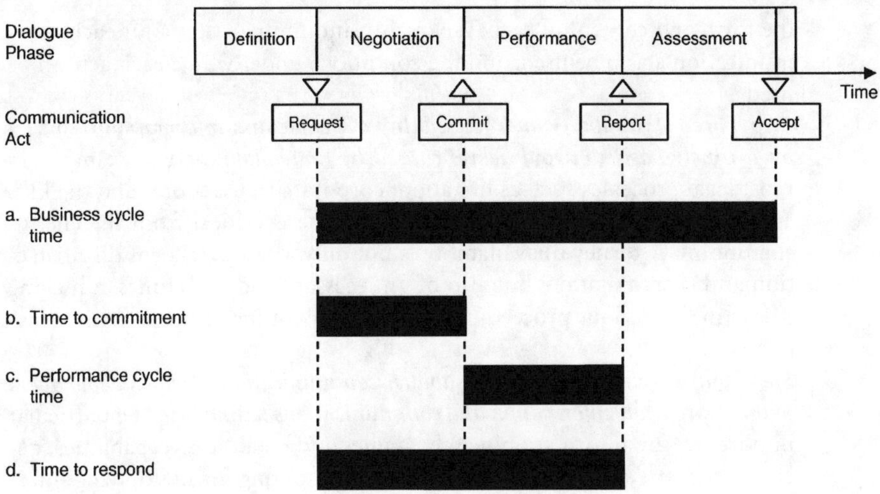

Figure 14.2. Cycle-Time Definitions

total of the time required to reach a commitment (reconfigure the commitment network) and the performance time needed to produce the newly committed outcome (Figure 14.2.d).

14.2.4 Time To Reach Commitment

As we said earlier, the time to reach commitment depends primarily on how quickly and effectively the negotiation phase of the customer/supplier dialog is conducted. The design approach includes several key principles applied in four steps:

1. *Step One: Initial process definition in terms of:*
   a. Process mission and goals
   b. Generic list of customers and process outputs (outcomes)
   c. Generic list of suppliers and process inputs (supplier deliverables)
   d. Environmental (fiscal, legal, regulatory, etc.) factors relevant to the process design
   e. Description of the primary customer/supplier relationship (between the business process and its immediate external customers), particularly important being specification of the related dialog in terms of the four phases (definition, negotiation, performance, and assessment) and seven possible communication acts
2. *Step Two: Creating continuous communication flow within all the customer/supplier dialogs that is characterized by minimum delay (waiting only for new information or knowledge — added value with no noise or redundant content).* This is accomplished by specifying use of only the seven communication acts;

the form and content of each; the quality and timing criteria for each communication act to be used; and the conditions under which each act can be used.

3. *Step Three. Availability and accessibility of people and information resources on a timely basis to avoid delays caused by unavailability.* For example, in most legal processes such as litigation, corporate transactions, and the like, access to the attorney responsible for the case is critical from the client's standpoint. Attorney unavailability is not only a cause of client dissatisfaction and even irritation, but also of unnecessary and sometimes damaging delays in the various proceedings, not to speak of the unnecessary increase in costs.
4. *Step Four. Accelerated customer-supplier communications through deployment of the appropriate information and communications technologies.* Of particular importance are information search engines and data access capabilities, decision-support systems, and collaborative computing for multiplying interactions among teams — which all are important tools in speeding up the flow of information.

Cycle time reduction achieved by applying these design principles to commitment management in the process will result in significant period performance improvements that relate directly to adaptability:

- Greater responsiveness to new or changing customer requirements
- Better ability to exploit small windows of opportunity
- Expedited rate of growth
- More effective application of information technology

14.2.5 Performance Cycle Time

Performance cycle time is determined primarily by how quickly and effectively the performance phase of the customer-supplier dialogue is conducted. The design approach essentially comprises the same key principles as described in Section 14.2.4; however, the following workflow-related considerations must be added:

1. *Step One.* Ensure that the procedural workflow does not include non-value-adding, repetitive, or redundant tasks; excessive inspections, measurements, and controls; or steps whose execution requires unusual preparations, expensive training, and complex technology. (Figure 14.3 shows the range of unnecessary or non-value-adding work steps.) Recognizing that speed is the result of simplicity (especially simple procedures and specifications), the creation of clear goal statements, process plans, guidelines, and protocols is key to good workflow design for cycle time reduction.
2. *Step Two.* Create continuous workflow within and between all the tasks of the process. In continuous workflow, everything (products, materials, energy, people, and information) moves continuously, stopping only when value is

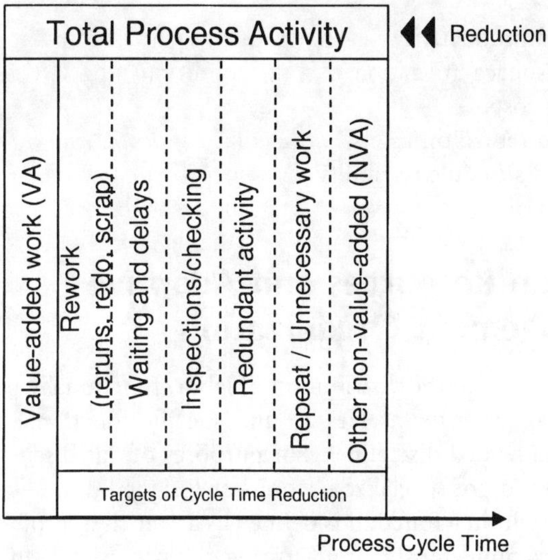

Figure 14.3. Process Cycle Time Reduction

directly being added or when loss of value (capital assets) must be prevented.[2] This is achieved by:

   a. Employing only the four value-adding work steps of *process, transport, inspect,* and *store*
   b. Establishing quality, efficiency, and cycle time criteria for each task
   c. Specifying measurements and controls at each hand-off point between tasks, and the conditions under which these may become choke points[3]
   d. Identifying the triggers that may initiate process improvement in the workflow (these triggers are the links to continuous process improvement, described in Chapter 16)

3. *Step Three.* Focus on the availability and accessibility of energy- and material-based resources (especially raw materials, energy, equipment, and other capital assets) on a timely basis, to avoid delays caused by waiting for a needed resource.
4. *Step Four.* Deploy the appropriate information and communications technologies to accelerate the flow of work by:
   a. Replacing or eliminating manual steps in the process
   b. Combining or integrating work steps
   c. Reducing the duration of individual work steps
   d. Enhancing resource availability or accessibility

Cycle time reduction achieved through these design methods in the process workflow will result in significant point-in-time (current) performance improvements directly related to effectiveness and efficiency and indirectly related to adaptability:

- Increased productivity
- Better resource utilization
- Reduced costs
- Increased return on assets
- Improved schedule reliability

14.3 Human Resources and Adaptive Management Organization

Based on the principle of the distribution of labor, functionally managed organizations typically configure themselves around specific functional areas and, within functions, around related disciplines. For purposes of our discussion, a function is defined as a collection of specialized actions or assigned duties within an organization. A discipline within a function is defined as a branch of professional learning or knowledge. For example, most manufacturing companies have an engineering function where most of the product-related design activities are carried out. Such important functions are often organized into divisions that in many corporations are the largest operating business units below the company level. Within the engineering division, engineering-related disciplines are grouped into smaller organizational units, usually departments such as mechanical engineering, electrical engineering, industrial engineering, and the like. Disciplines may be further divided into subgroups within a department, configured around still narrower, discipline-related specialties. For instance, within an electrical engineering department, we may find sections for power engineering, control engineering, transmission engineering, etc.

As a consequence of this functional approach, we end up with organizations comprising groups that represent functions, disciplines, and subdisciplines which are representations of expertise, rather than outcome. The people identified with the various disciplines and subdisciplines typically reside in a hierarchical, command and control management structure also referred to as "silos."

This silo approach to organizational design results in work being completed in functional or discipline-defined chunks that are often handed off to another silo for further processing until the complete work assignment is accomplished across multiple functions. As we have shown earlier, this arrangement requires an excessive amount of coordination, especially at the hand-off points between the silos, where the requirements of individual silos are largely defined by the internal rules and work practices within the silo, rather than the ultimate customer for whom the entire work effort is undertaken.

As a result, such a narrowly restricted and disjointed approach to work will almost always be accompanied by inefficiencies, as well as quality and schedule-related problems. For this reason, traditional process reengineering saw its greatest opportunity in the rearrangement and elimination of hand-off points, those interfaces in the process where an interim work product is being passed from one organizational unit (division, department, or section) to another.

The interfaces we have described so far, however, are not just the hand-off points where interim work products are passed from one function to another. Many interfaces are also singularities, where intensive sets of communications take place between the affected parties. This information exchange is part of the ongoing dialogue which takes place at all key interfaces within the process as well as at the interfaces of the process with its external environment. It is our view that the holistic approach to process management can deal more effectively with both challenges: the elimination of unnecessary hand-off points and the redesign of functional interfaces into the new reality of customer-supplier dialogues.

Our approach is based on the premise that organizations must behave as adaptive systems — where adaptability is supplied by individual, key business processes. The role of the CEO and the senior management team is to act as the transformation architects of the corporation. They can usually do this in their respective roles of transformation executive and transformation council. In assuming these roles, management is not concerned with the details of company life, but with selecting and focusing on the correct processes at the right time to enable the rest of the key processes to interact optimally. In other words, it is senior management's task to create a unique transformation architecture that, when implemented, will result in alignment around key business processes — not to design or reengineer individual business processes within the organization.

Accordingly, the CEO as a transformation leader should develop a vision while allowing each business process owner and team to develop its own process goals and objectives consistent with that vision. Each CEO should select and name the business process owners but not try to determine the composition of individual process management teams. The CEO should design the transformation process but not get involved with the design or operation of the individual business processes.

At the process and subprocess level, empowered and accountable process ownership is the organizational mechanism designed to ensure continuing, coherent performance focus on the entire business process. Such undivided attention to the business process is a prerequisite for ensuring process adaptability. Accordingly, the key responsibilities of the process owners include:

1. *Commitment management*, which is
   a. The negotiation and agreement on the commitment needed to satisfy the customer's requirements
   b. The continuous tracking and updating of commitments to ensure satisfactory response to changing requirements

   The most important result of good commitment management is process adaptation. This is the continuing enhancement of process capability to ensure effectiveness and efficiency in the face of changing or new requirements, both expected and unexpected, over time. To achieve this from an organizational standpoint, it is mandatory to empower subprocess owners and their teams to sense and evaluate changes as they occur and to identify specific capability enhancements in response. These enhancements would include:

   a. Means of eliminating or circumventing existing constraints
   b. Identified opportunities for improving specific activities
   c. Alternative sources of resources to carry them out (internal processing vs. outsourcing)
2. *Performance management through process control*, which requires maintaining workflow stability for the duration of the performance phase to meet agreed-to conditions of customer satisfaction and thereby ensure process effectiveness
3. *Process optimization*, which deals with maintaining the process in a state of efficiency until customer requirements are met; for purposes of this discussion, we define process optimization as the management actions needed to simultaneously maintain process effectiveness and efficiency over a finite period of time

The important underlying fact of the organization of human resources in the holistic process environment is that structure and reporting lines have a declining significance and the basic relationship between employer and employee is changing. Process centering requires that individuals think differently about their work, roles, relationships, and the chain of command. Processes and organizations are transformed to become more adaptive, but also to change from loyalty (expected by the employer) and security (expected by the employee) to a relationship of shared objectives, knowledge, and rewards.

14.4 Information Technology

In Chapter 10, we discussed the basic attributes of the information system architecture best suited as an enabler and integrator of effective response to change both at the enterprise and individual business process level. We also introduced the key design and operational characteristics of collaborative systems. These are particularly favored as the networked information management and computing environment in which the high rate of customer-supplier interactions and information-intensive knowledge work can best be carried out.

At the business process level, the ideal information management environment is one where empowered and accountable people, aided by collaborative technology, manage the customer-supplier dialogue and carry out all related communications. This is accomplished while automated application directors, workflow managers, network controllers, and database administrators ensure the effective and efficient execution of all procedural work activities and the delivery of the outcome.

The information flow in such a business process essentially contains: (1) the communications between customer and supplier organized around the seven communication acts, and (2) descriptive information associated with work activities and their performance and products (essentially, the information used or generated by work activities, as shown in Figure 14.4).

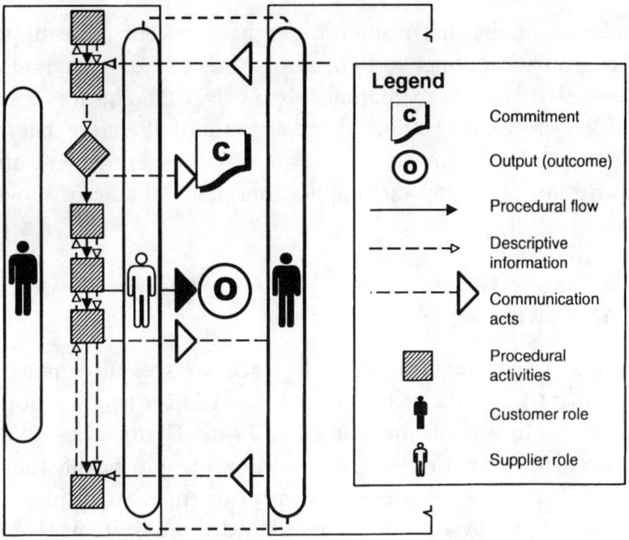

Figure 14.4. Descriptive Information Flow and Communication Acts

The basic systems and communications technologies supporting the process reengineering and process management environments have seen significant and rapid change over the past decade. A 1993 survey[4] of Fortune 500 companies on the impact of technology on reengineering efforts shows the top six technologies, headed by desktop computers. A more recent study[5] shows a very different top six technologies, headed by wireless communications (Table 14.1).

Table 14.1. Relative Impact of Technology on Business Process Reengineering

Top Six Technologies (in order of decreasing impact)	
1993	*1998*
Desktop computers	Wireless communications
Electronic data interchange (EDI)	Laptop computers
Shared database	Local area/wide area networks (LAN/WAN), Internet, World-Wide Web
Expert systems	Shared database
Local area/wide area networks (LAN/WAN)	Messaging systems (e-mail)
Optical scanning	Collaborative computing

Our focus here is on the type of applications that are particularly supportive of the tracking and coordination of networks of commitments and the myriad subprocess-level activities they drive. Given that all this involves the continuous and intense interaction of definable, but constantly changing groups, it is clear that the information systems designed to support the holistic process management approach fall within the domain of (1) groupware applications, and (2) enterprise-wide applications systems.

14.4.1 Groupware

As the design and management of business processes is usually carried out by the process management team, the value and utility of team-oriented groupware applications should be intuitively obvious. Based on Esther Dyson's view of the different types of groupware information systems,[6] it is convenient to classify these along two intersecting lines as process-oriented and control-oriented groupware.

Process-oriented groupware is further subdivided into systems that (1) support the input-process-output type of descriptive information flow representing such items as actions and data, and (2) support and track the flow of information representing communication acts and, in particular, commitments. Examples of the former can be found in manufacturing industries. Bar-code scanners and their supporting software help to monitor the elapsed time in each process step from initial assembly all the way through shipment and delivery. An example for commitment-oriented software applications is online electronic trading, where the entire transaction is based on a series of requests, offers, and commitments.

Control-oriented groupware is designed to support the various levels of control within the process. It can be further subdivided into user-oriented systems, workflow-oriented systems, and outcome-oriented systems.

User-oriented systems support individual end users manipulating, transforming, and controlling data or using data to interact with others. Examples are messaging systems such as Lotus Notes, electronic data interchange (EDI), data mining and warehousing applications, network search engines, and many online professional and management tools such as process modelers.

Workflow-oriented systems support the conversion or movement of data to be attached to work products for identification, measurement, or control purposes (electronic forms, online imaging, and certain multimedia applications fall into this category).

Outcome-oriented systems support process management by tracking and linking work products and related communications between customers and suppliers; collaborative project management, workflow management, and scheduling systems belong to this category (for example, applications which monitor whether a customer has accepted a shipment reported to have been made by the supplier). Of these, a combination of information flow-oriented process software and outcome-oriented control groupware is the ideal system environment.

14.4.2 Enterprise Applications

The challenge with today's large variety of process-related software applications, such as process mapping or modeling tools or workflow managers, is that they are mostly designed to provide a particular solution for a particular type of process problem. Messaging software has become popular because it is one of the few cross-functional software packages on the market that has integrated a number of end-user functions into one large communication-based database management system. Many packages on the market, however, cannot communicate well with each other and for this reason cannot seamlessly cover the three functional categories of process management applications:

1. Process mapping and modeling (descriptive or diagramming techniques)
2. Process implementation planning and deployment (project management tools)
3. Operational process management (simulators, workflow and information flow managers, collaborative project managers, etc.)

The advent of enterprise-wide application systems (also dubbed "wall-to-wall software" but better known by the correct technical term of enterprise resource planning, or ERP, systems), promises to address this dilemma because of these key systems attributes:

- *Multifunctional scope:* The system offers the ability to track a variety of data types across the enterprise, such as costs and other financial results, materials, people, sales, and inventories.
- *Integrated information change management:* Entry of any new or changed data causes the automatic change/updating of all other related data across all key functional areas (such as in manufacturing, customer-facing areas, and the "back office").
- *Modular and scaleable structure:* The system can be implemented on as narrow or as broad a base as one chooses, in a small local company or in a global enterprise.
- *Platform independence:* ERP systems will operate both in a mainframe or client-server environment.

Enterprise resource planning systems should be implemented for good strategic business reasons, such as integrating and supporting a business process reengineering or process-centering effort or as part of a comprehensive, long-range information technology strategy. An ERP implementation should never be embarked on for purely technological reasons or to solve chronic functional or organizational problems. Similarly, the ERP software vendor should be carefully chosen, consistent with the industry the enterprise is in, the size and complexity of the business, the scope of functionality needed, and any other unique process-related requirements.[7]

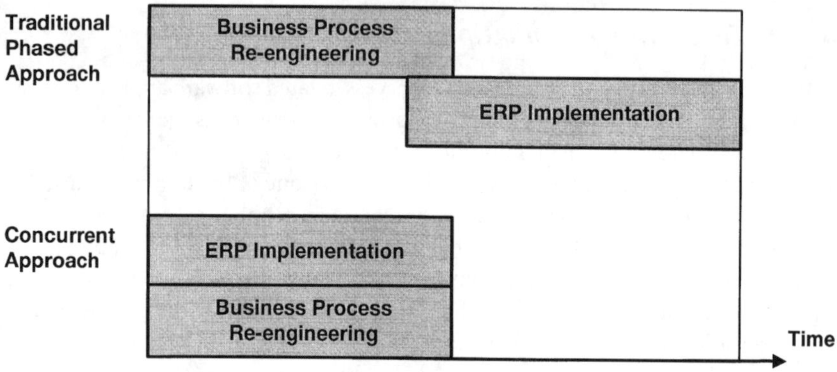

Figure 14.5. Business Process Reengineering with ERP

transformation, in contrast to traditional reengineering, which is aimed only at current performance improvement. For this reason, fundamental reengineering cannot be accomplished after an ERP solution has been decided on; the only two choices available to management are either to (1) reengineer the business processes first and then design the ERP solution afterward, or (2) reengineer the business processes and develop the ERP solution concurrently. These two implementation approaches are illustrated in Figure 14.5.

Because most ERP methodologies are also designed to be roadmaps for the implementation effort, the second choice is less risky and its implementation has, so far, shown a higher rate of on-time and on-budget success. An example for this approach is the reengineering of the order-fulfillment process (OFP) in Union Camp's Fine Paper Division. When Union Camp embarked on implementing its ERP system solution, the overriding business reason was to enhance, right from the outset, the process reengineering effort with the latest and most effective available applications technology.

Union Camp chose Baan as the ERP solution provider based primarily on several attributes of the Baan system and implementation:

- Recognized process industry expertise
- Process design for fulfilling orders
- Fit of production-related modules with the Union Camp mill operation
- High level of compatibility between process modeling modules and the reengineering methodology[8] employed by Union Camp
- The strong link between design and implementation provided by the Baan system

The detail-level process design and Baan system design have been proceeding hand-in-hand and practically in parallel; in other words, the OFP process management team and their Information Technology (I/T) colleagues have chosen the

- Fit of production-related modules with the Union Camp mill operation
- High level of compatibility between process modeling modules and the reengineering methodology[8] employed by Union Camp
- The strong link between design and implementation provided by the Baan system

The detail-level process design and Baan system design have been proceeding hand-in-hand and practically in parallel; in other words, the OFP process management team and their information technology (I/T) colleagues have chosen the concurrent implementation alternative. The full reengineering effort started in 1996, and the ERP decision was made in the summer of 1997. The concurrent Baan applications system implementation was running on schedule and within budget as of late 1998, with its completion planned for a year later.

Regarding the use of ERP solutions in conjunction with the holistic process design and process centering, a few evolving aspects should be taken into account. First, most process mapping and modeling software applications, whether used as stand-alone tools or with an intended link to an ERP system, are based on the input-process-output process model. There are a few exceptions,[9] but the better known process modelers that can be linked to ERP solutions (e.g., SAP, Oracle, or Baan) will only handle procedural flows.[10] Second, while most ERP systems now include their own process description (mapping and modeling) tools, these are based on the process design as determined by the ERP system's software designers; however, their "ideal" design may or may not be fully compatible with what many companies need. One of the challenges for ERP software vendors is to include in their applications systems the modeling capabilities that support the holistic process definition presented here.

14.4.3 Business Processes as Assets

The traditional accounting view that treats business processes as sources of expenses can lead to unexpected problems. Companies can seriously underestimate the true costs of their processes in terms of the total amount of capital assets tied up in deploying a process and its ongoing operation. As a result, a corporation may assume that a particular business process performs well when it actually costs more in capital than it contributes to cash flow. The often-heard assertion that massive corporate investment in information technology over the past three decades has not resulted in measurable bottomline improvement is just one symptom of this problem. Also, viewing business processes as mere expenses ignores the fact that processes generate learning and knowledge that can be turned into intellectual or capital assets of considerable future value. The inclusion of information flow and knowledge into the process model and, by extension, the information systems that manage and integrate them also allows management to view business processes not as expenses, but as enterprise assets.

14.5 Summary

All work is done in processes; therefore, performance measurements are most meaningful when applied to the process and its products. Process performance measurements cover both dimensions of managing commitments and managing the procedural workflow.

Quality controls are still applied to the procedural flow to ensure customer satisfaction; however, they play a much more important role as time controls for decision and commitment. Another important category of quality controls ensures the completeness of content, the clarity, accuracy, and precision of the communication acts within the customer-supplier dialogue. Cycle time reduction, particularly with respect to the time to respond, is a key indicator of process adaptability.

The concept of the adaptive management organization is based on the premise that organizations must behave as adaptive systems in which adaptability is supplied by individual, key business processes. The holistic approach to process definition and management requires a human resource organization based on empowered and accountable ownership.

Information technology is designed to accelerate and enhance both communications and the flow of work in the process. In this respect, the value and utility of team-oriented groupware applications are obvious. Information technology also has uniquely tailored capabilities to integrate, within the process and across the enterprise, the management of commitments and outcomes. The advent of enterprise-wide application systems, known as enterprise resource planning systems, promises to address the seamless integration of process planning and implementation technologies.

15 Design for Adaptability

The holistic process model provides the base for designing a highly adaptable and responsive business process. Adaptability is a process attribute that must be carefully designed into the business process through (1) robustness of the backbone network of commitments, (2) features resulting in reduced cycle time and structural flexibility to accommodate needed incremental changes, and (3) improvements in workflow or resources.

Every process must be designed — or redesigned — around four basic *design concepts:*

1. Uniqueness of purpose
2. Customer-supplier value chain
3. System of feedback
4. Complete set of resource-consuming assets

Of these, the customer-supplier value chain and the system of feedback provide the basic criteria for process adaptability.

Traditional companies typically struggle with their environment, seeing it as a source of disruption, unwelcome change, and negative influence to their organization. These companies tend to isolate themselves from their environment as long as possible, hoping to preserve the state of stability they think they have attained. They will, at most, respond to forces and changes outside their boundaries on a selective basis while still maintaining a predominately controlling, defensive posture. The result is an inherent tension between stability and openness, constraint and flexibility, control and adaptation.

Over time, many organizations, and even societies, have experienced decline and even demise because of self-imposed, defensive isolation from their environment.

The best recent historical example of this phenomenon in the extreme and its ultimate consequences is the history of the Soviet Union and the Communist social order. The Communists were paranoid not only about foreign spies and saboteurs, but also about the entry of any information into their society that did not pass through the censure of the state. The results were an ossified political structure, a society devoid of creativity and innovation, and an economic system without incentives and value creation that eventually doomed it to failure and collapse.

In self-renewing, adaptable processes, these new realities manifest themselves in many different ways. This chapter looks at structures that are capable of maintaining their functional identity (what they do) while changing their form (how they are doing it). Neither form nor function alone dictates how the process is configured. Instead, form and function interact in an elastic way where the process may maintain itself in its current form or evolve to a new, improved level. The key is that the process is capable of spontaneously creating new structures depending on what is required. This means that it first organizes information (commitments) in the form that best suits the current need and then uses that information to define the required process flow and the resource-consuming assets for executing it. Every time the need changes, commitments are reorganized and process flow and resources redefined.

This type of adaptive behavior in processes is critical for the responsiveness and competitive survival of the enterprise of which they are part. Haeckel correctly concludes that in an environment of unpredictable change, adaptability is of greater value than efficiency.[1] Companies that have reengineered their key business processes for better efficiency in the traditional way have learned this lesson the hard way when they realized that those same processes were unable to adapt to unexpected change — another case of the process paradox.

15.1 Traditional Design Approach to Adaptability

Traditional process reengineering based on the input-process-output model cannot address satisfactorily the problem of changing customer requirements. Because this model is essentially procedural in outlook, the approach taken to the problem is also procedural. Indeed, the most frequent way out of this dilemma is the introduction of a customer-supplier interface known as the "gearbox," first introduced by Juran Institute in 1995.[2] As illustrated in Figure 15.1, this interface comprises the information flow surrounding the delivery of process output to the customer. In this representation, the information flow across the interface is divided into two categories. The first category deals with the information related to the workflow and its product — descriptive information accompanying the product to the customer and feedback about the product flowing from the supplier to the customer. The second category deals with bi-directional communications between the customer and supplier about changed or future customer needs.

To handle this information flow, the traditional approach creates a procedure within the business process known as the customer requirements generation subprocess, or simply gearbox subprocess (see Figures 11.4, 12.5, and 15.2). Similarly, a

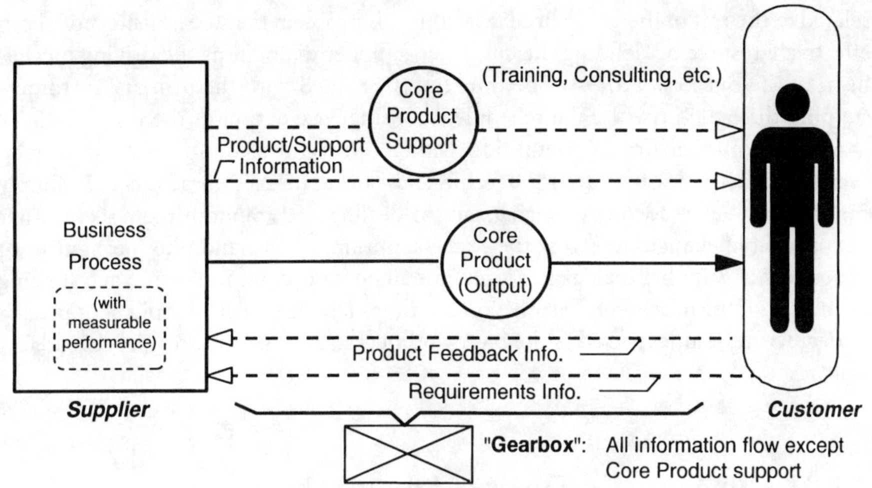

Figure 15.1. Customer-Supplier Interface

corresponding activity is inserted into each subprocess, a similar task into each activity, and so forth — in recognition of the fact that this type of interface must exist at every hand-off point and at every level of the process.

What the traditional approach has difficulty managing, however, is the timing and sequence of unscheduled communications between customer and supplier, as

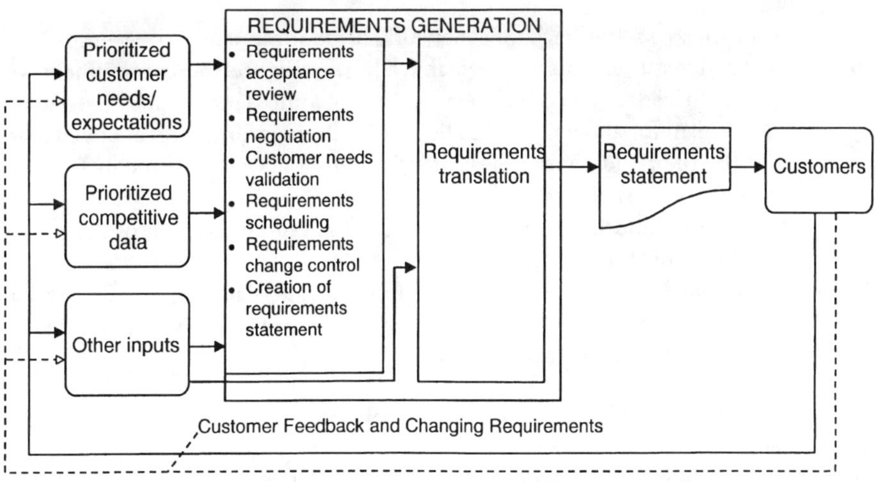

Figure 15.2. Customer Requirements

related to the rest of the procedural flow. It is this problem that the holistic model can effectively resolve by defining the customer-supplier dialogue as the driving mechanism of the business process. The time line associated with the primary customer-supplier dialogue provides the schedule for a single execution of the process.

The traditional process definition with its procedural outlook can only map events and activities that are foreseeable and can be defined in advance. It simply cannot provide an adequate mechanism for dealing with random requests or other unpredictable changes affecting the process. On the other hand, the mechanism of the customer-supplier dialogue provides for all possible communication acts flowing in and out of the process and at any point in time. This ability of the holistic approach to deal with the unexpected is the basic and necessary condition for process adaptability.

15.2 The Holistic Design Approach

As defined earlier, adaptability is the inherent ability of the business process to recognize and respond to change while maintaining its own effectiveness and efficiency over time. The required speed of response or the magnitude of change being responded to are critical ingredients of process adaptability. A process that can absorb bigger change, or respond to the same change faster than other similar processes, is said to be more adaptable and, consequently, more competitive.

Adaptability is a process attribute that is carefully designed into the business process by addressing both dimensions of the process: commitments and workflow. As most existing process designs do not include the commitment component, adaptability can only be achieved through true process transformation, also known as fundamental reengineering. Process adaptability has three key features:

1. Robustness in the backbone network of commitments to accommodate changed requirements and dynamically reconfigure the commitment sequence to lead to the desired new process outcome. This includes an effective mechanism for allowing (a) the flow of feedback information about change, and (b) the rapid communication of resulting new commitments to guide the flow of work activities.
2. Continuous and fast workflow to ensure the rapid execution of the procedural part of the response.
3. Structural flexibility in the process comprising alternate paths of work activities to accommodate (a) needed changes and improvements in the workflow, or (b) the amount of resources used.

Alternate paths or resources may mean outsourcing the needed procedural activities for better efficiency or shorter cycle time — that is, quicker response. The design approach to adaptability varies, depending on whether a new business process has to be designed or an existing process reengineered. However, both approaches are guided by a set of basic process design concepts.

15.2.1 Process Design Concepts

The term "process design" as used in this text refers to the activities of planning the business process in order to achieve its stated objectives — that is, to meet customer requirements within agreed-to conditions of satisfaction. There is no such thing as process design in the abstract; the related planning is always done to meet known (current or anticipated) requirements. The end result of the process design cycle is the *process plan*; it may endure over a long period of time or become subject to change very soon, depending on the dynamics of the surrounding environment.

In general terms, the process plan describes how the process is supposed to operate and how its performance is evaluated and controlled. The most critical components of the process plan include:

- Communication protocol for negotiating commitments
- Conditions of satisfaction
- Criteria for renegotiating commitments and introducing change into the business process

The process plan also includes the measurement and control plans for both the commitment and workflow dimensions. In short, the process plan is the process owner's blueprint for ongoing process management and process change (Table 15.1).

As already stated, every business process — technological or administrative, production or service oriented, complex or simple — must be designed around these four fundamental design concepts:

1. It has a *uniqueness of purpose* (also known as process mission), which defines the process in a singular manner and differentiates it from other business processes within the organization. The basic characteristics of the unique business process include the following:
   a. The process must be capable of producing the committed outcome and be in control while actually executing work activities so that it is effective in meeting the stated conditions of satisfaction. Process effectiveness is one necessary precondition to adaptability; it is ensured through continuing process improvement.
   b. The process must be efficient; it has to operate at a minimum level of resource consumption. Process efficiency is the other necessary precondition to adaptability, and it is ensured through process optimization.
   c. The process must be adaptable; there has to be a predefined but flexible way of changing it over time as a result of variations in output or by changing customer requirements. Only an adaptable process can absorb change without adverse impact on its effectiveness and efficiency.
2. The customer-supplier value chain is comprised of the set of customer-supplier relationships and related dialogues.
3. A basic *system of feedback* of control loops ensures that those responsible for managing the various activities are always in a position to know what is to be done, as defined by existing commitments. They must also know what is

Table 15.1. Process Plan

Process Name	Process Owner	Date Developed/ Date Revised
Process mission		Developed: _______ Revised: _________
Process goals		Developed: _______ Revised: _________
Process management infrastructure Owner Team Stakeholders		Developed: _______ Revised: _________
External process customer(s)		Developed: _______ Revised: _________
External customer needs/requests		Developed: _______ Revised: _________
Internal customer-supplier value chain Customer-supplier relationships Customer-supplier dialogue definition		Developed: _______ Revised: _________
External communication protocol		Developed: _______ Revised: _________
External customer conditions of satisfaction		Developed: _______ Revised: _________
Commitment renegotiation criteria		Developed: _______ Revised: _________
Process outputs		Developed: _______ Revised: _________
External process supplier(s)		Developed: _______ Revised: _________
Process input(s)		Developed: _______ Revised: _________
Process map		Developed: _______ Revised: _________
Process measurement plan		Developed: _______ Revised: _________
Process control plan		Developed: _______ Revised: _________
Process resource summary		Developed: _______ Revised: _________

being done, so that corrective action can be taken if there is a discrepancy between the two. Without the basic control loop, no set of activities and no collection of resources, however well chosen and ordered, will ensure effective operation and conformance to the conditions of satisfaction. Learning loops within the customer-supplier dialogue and along the value chain help monitor and coordinate commitments and customer assessment of outcomes. Adaptive loops across business processes keep track of the backbone network of commitments and help manage changes in commitment content and network configuration.

4. There is a clear and complete definition of *resource-consuming assets*: people, equipment, materials, energy, information, knowledge, and methods.

Incorporation of all four concepts is necessary for good process design and satisfactory process operations. Whether the process is managed and controlled by sophisticated methods and technologies or by simple visual observation and human intervention, these four design concepts will determine its ultimate success or failure.

15.2.2 Design of New Process Structure

The definition of a holistic process structure seems to be the easiest when no previous process exists or when existing, stand-alone process capabilities (subprocesses) have to be assembled into a new, cohesive, working business process. The steps of this type of original process definition are illustrated by the circled numbers in Figure 15.3:

1. Identify the external customers of the business process. This establishes the primary customer-supplier relationships between the process and its customers.
2. Determine either (a) customer requirements (the requests made by the customers upon the process), or (b) potential process obligations as the result of offers made to one or more customers.
3. Define the customer-supplier value chain within the process, for each external customer and for each type of customer request received (or offer made). This is best done by tracing the original request along the chain of all involved parties within the business process. This approach will yield the flow of subordinate requests being made within each customer-supplier relationship in the chain.
4. Complete the customer-supplier dialogue for each empowered customer-supplier pair in the chain by defining all seven possible communication acts (request or offer, commit, report completion, accept, reject, or withdraw) within each dialogue — with special emphasis on the agreed-upon commitment and the related outcome.
5. Define procedural activities (workflow) and the related descriptive information flow required to produce the committed outcome in each customer-supplier relationship along the chain. Most of these activities are carried out in the performance phase of the dialogue. The descriptive information is data

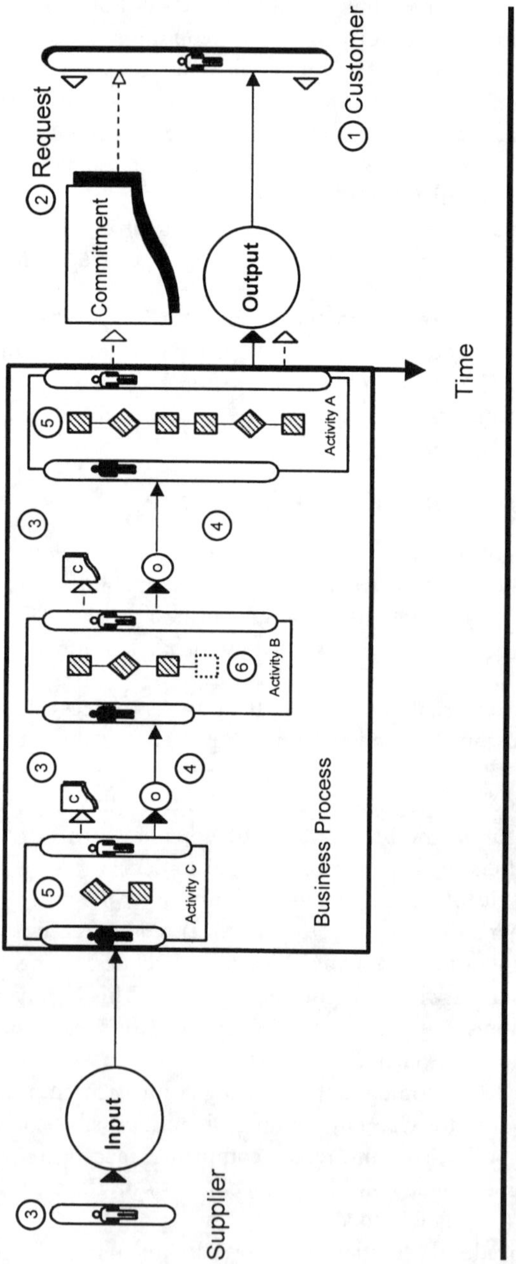

Figure 15.3. Steps of Original Process Definition

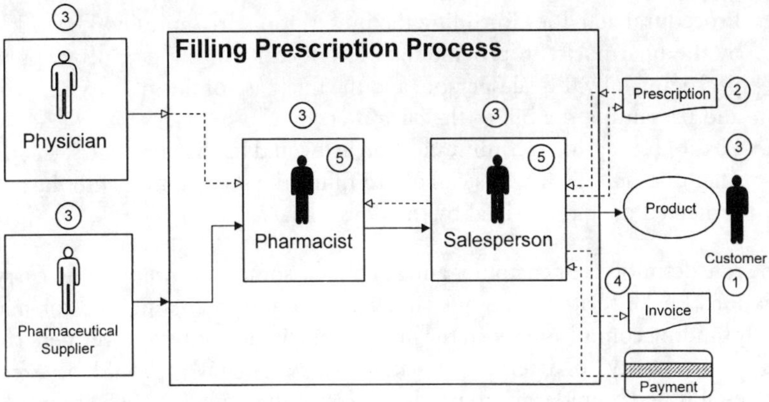

Figure 15.4. Pharmacy Operation

used or generated by each procedural activity or task in the workflow. Most of the data will relate to and accompany the work products passed between activities. Some data, however, will cross the boundaries of the business process as, for instance, control information of value to other business processes within the enterprise. (It should be noted that this is the type of information picked up by inter-process learning loops or adaptive loops.)

6. Define procedural activities and the related descriptive information flow supporting all other possible communication acts (counter-offer, withdrawal, and so on) that are treated as exceptions to the normal sequence of events in each customer-supplier relationship.

To illustrate the approach to original process definition, let's consider the case of an ordinary pharmacy (Figure 15.4). The business process to be defined is the filling of prescriptions. The six steps of process definition would be as follows:

1. For this discussion, external customers of the business process are defined as being those who bring in a new prescription to be filled or have one already on file and wish to have it refilled.
2. Customer requests are to fill their prescriptions.
3. The value chain for filling prescriptions includes the customer, the salesperson, the pharmacist, the physician who provided the prescription, and the supplier of pharmaceutical products.
4. The customer-supplier dialogue between the customer and the salesperson, for example, includes the customer's request, the salesperson's commitment to have the prescription filled by the pharmacist, the salesperson producing the prescribed product, its availability, the printing of an invoice, and finally the customer's acceptance of the product through payment of the bill with cash or a credit card.

5. Procedural activities for filling the prescription are typically the work done by the pharmacist to provide the prescribed item, the task of wrapping or packaging it by the salesperson, and the final task of the salesperson printing and presenting the bill to the customer.
6. Possible exceptional communication act would be the pharmacist's counter-offer of a generic drug instead of just filling the prescription with the patent-protected drug prescribed by the physician.

The role definitions of customer and supplier assume empowerment to carry out negotiations, make related decisions, and arrive at agreements needed for making mutually binding commitments. In the procedural dimension, empowerment simply means the authority to determine what activities and tasks should be executed (within established boundaries) to produce a committed outcome. The supplier role also establishes the accountability for the outcome committed.

15.2.3 Redesign of Existing Process Structure

The redesign of an existing business process is preceded by a comprehensive analysis of customer requirements and a data-driven analysis of current process performance, and its history, if any, of past responsiveness. The basic purpose of this analysis is to determine whether the business process has been defined and operated purely as a procedural sequence of repeatable activities. In such a procedural workflow, the information content is only a consequence, not the driving force. Such business processes may have been in existence and operational for some time. Therefore, the purpose of the process design effort is twofold:

1. To establish a new dialogue and value-chain structure by designing and then implementing communication protocols, the network of commitments, and commitment management.
2. To upgrade and improve the workflow and its performance in terms of effectiveness, efficiency, and cycle time.

When the process designer faces the task of upgrading an already existing procedural process representation (IPO model) to a holistic definition, the sequence of definitional steps is somewhat different (Figure 15.5):

1. Identify and assign procedural activities and the related descriptive information flow to the parties (individuals, teams, departments, and so on) performing them. This will essentially define the structure of the customer-supplier value chain.
2. Establish each customer-supplier relationship by first identifying all empowered customer-supplier pairs along the chain and then adding the appropriate customer-supplier dialogue to each. This essentially defines the content of the value chain and most importantly, the empowered individuals in it.

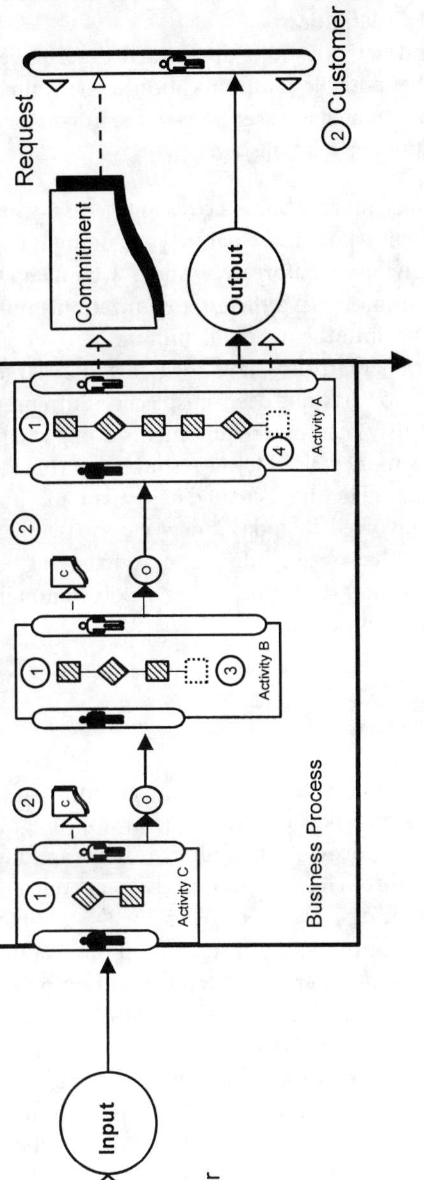

Figure 15.5. Steps of Existing Procedure Redefinition

3. Retrofit, as appropriate, the procedural workflow with activities and information flow resulting from the newly established dialogues not provided for in the original IPO definition.
4. Define procedural activities and the related descriptive information flow supporting all other possible communication acts (counter-offer, withdrawal, and so on) that are treated as exceptions to the normal sequence of events in each customer-supplier relationship.

Process design or redesign is not easy but is a constant necessity in the face of change. Having process professionals who know what they are doing is essential. First efforts may not be perfect, and giving up before everyone is acclimated to the discipline of process design is not uncommon; nevertheless, organizations must reconfigure their processes or not survive far into the new millennium.

Nimble, entrepreneurial enterprises have been envied by larger organizations in recent years. That admiration has not been misplaced. Entrepreneurial enterprises have figured out how to satisfy customer requirements doing whatever it takes. These enterprises have spent less money, used fewer resources, involved fewer people, and done the work in record time. They have created new, adaptable processes that get the work done. The people involved in today's new enterprises have abandoned the structures and procedures that once made corporations great because they realize that those structures and processes are no longer flexible enough. So, learning the basics of process behavior is essential.

15.3 Summary

In an environment of unpredictable change, adaptability is of greater value than efficiency. The holistic process model provides the base for designing a highly adaptable and responsive business process. Adaptability must be carefully designed into the business process primarily through robustness of the backbone network of commitments. Every process must be designed — or redesigned — around four basic design concepts: (1) uniqueness of purpose, (2) customer-supplier value chain, (3) system of feedback, and (4) completeness of resource-consuming components. Of these, the customer-supplier value chain and the system of feedback provide the basis for process adaptability. In addition, the design for adaptability includes features that help reduce cycle time and the structural flexibility necessary to accommodate needed incremental changes and improvements in workflow or resources.

Traditional process reengineering based on the IPO model cannot satisfactorily address the problem of changing customer requirements. It is this problem that the holistic model can effectively resolve by defining the customer-supplier dialogue as the driving mechanism of the business process. The dialogue mechanism provides for all possible communication acts flowing in and out of the process and at any point in time. This ability of the holistic approach to deal with the unexpected is the basic and necessary condition for process adaptability.

The end result of the process design cycle is the process plan. The most critical components of the process plan are the communication protocol for negotiating commitments, the conditions of satisfaction, and the criteria for renegotiating commitments and introducing change into the business process. The process plan also includes the measurement control plans for both the commitment and workflow dimensions.

The definition of a holistic process structure seems to be the easiest when no previous process exists, or when existing, stand-alone process capabilities have to be assembled into a new, cohesive, working business process. When the process designer faces the task of upgrading an already existing procedural process representation (IPO model) to a holistic definition, the sequence of definitional steps is somewhat simpler; however, the definition phase is preceded by a comprehensive analysis of customer requirements and a data-driven analysis of current process performance, and its history of past responsiveness.

16 Design for Robust Commitments

Robustness of commitments means that the following are known:

- What commitments exist
- Which ones need to be changed
- What new commitments need to be made to accommodate new situations
- What decisions have to be made
- Who is going to make this all happen

16.1 Design for Process Adaptability

As we have said before, the holistic process model is the basic design structure for adaptability. Once the process structure has been defined, the design for adaptability is carried out in three steps.

1. *Step One.* Completely describe the backbone network of commitments connected by learning loops attached to the communication acts within customer-supplier relationships (Figure 16.1).
2. *Step Two.* Establish the reconfigurability of this network and the work activities it drives by defining and linking specific capabilities of the workflow to the fulfillment of specific commitments. This linkage of workflow and commitments requires detailed process plans right down to the task and work step level.
3. *Step Three.* Ensure robustness through the clear identification and assignment of accountabilities throughout the process. This will ensure that the path of decision-making is clear and that those who will decide about changing commitments are empowered and able to do so.

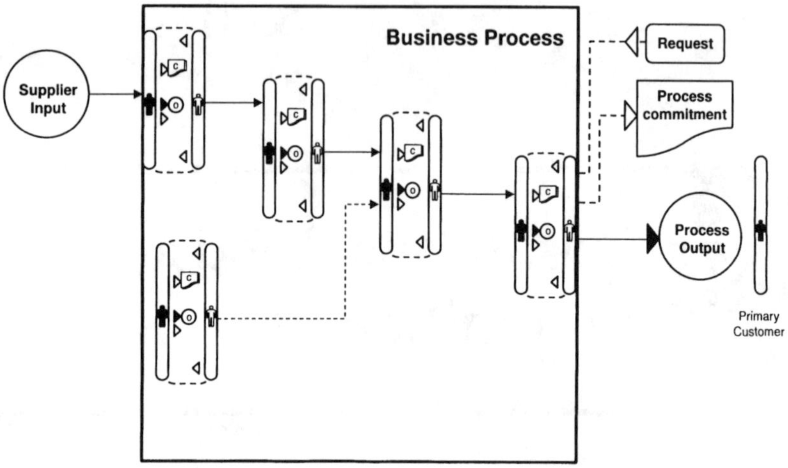

Figure 16.1. Backbone Network of Commitment Management

16.1.1 Backbone Network of Commitments

Within a given business process, the backbone network of commitments is defined as that set of commitments along the customer-supplier value chain which is critical for satisfying a new or changed requirement in the primary customer-supplier relationship (Figure 16.1). A trip on an airplane is a prime example of how commitments can make or break its success. By the time customers board a plane, airline reservation personnel, travel agents, or computer website transactions have handled the reservation process. The airport has made sure that there is parking, check-in desks, baggage handlers, etc. The airline has scheduled a flight crew. A clean-up crew has prepared the plane for the passengers. A caterer has loaded its food on the plane. Many other tasks have been completed as well.

It is this set of commitments that may have to be renegotiated to achieve a new outcome. For example, if an airline cannot get a passenger to his or her destination, existing commitments may need to be changed. This changed set of commitments will ensure a new outcome: later arrival at the destination, for example. If this outcome is agreeable to the customer, then a seat has to be found on another flight and the passenger's baggage transferred to the appropriate plane. The flight must depart and arrive on time, an arrival gate must be available, and so on until a satisfactory outcome is achieved. (The converse of this situation, however, is not true: according to the principle of equifinality,[1] the same outcome may be achieved by following different interim paths of commitments and actions.)

Generally speaking, not all commitments along a given value chain are critical. In fact, when the original chain of commitments is established to support the primary commitment, the critical ones are linked by a special type of information flow known as learning loops, as tangible recognition of their interdependence. These learning loops will, in turn, provide information for those accountable for the satisfactory fulfillment of the related commitments.

16.1.2 Workflow Reconfiguration

Work in a business process is always performed in the context of one or more customer-supplier relationships along the value chain and always in support of a committed outcome. When commitments change or new commitments must be made, it is essential that the required process capabilities are easily identified and swiftly deployed for execution. Therefore, the workflow design must systematically link in detail the procedural aspects of the process to the customer-supplier relationships they support. This systematic linkage is accomplished by disciplined, methodical process planning, for *there is no commitment without a plan.*

As we have said earlier, for detailed planning it is convenient to break down a large business process into subprocesses, subprocesses into activities, activities into tasks, and tasks into individual work steps. These definitions of workflow segments assume additional meaning and significance in the holistic business process structure. They come to represent work capabilities that can be purposefully identified, arranged, or reconfigured to fit the four phases of the customer-supplier dialogue and the requirements flowing from them (Figure 16.2).

In the holistic process structure, a *subprocess* is the sequence of activities required to support the fulfillment of a commitment and the related completion of all four phases of the customer-supplier dialogue. In other words, a subprocess represents the work to be performed to complete a single transaction or a business cycle. (The reader is reminded that a given business process may be involved in several transactions concurrently, each of which may represent a different subprocess in execution.)

An *activity* is the smallest unit of work required for the completion of one phase of the customer-supplier dialogue resulting in a specific communication act, such as making a request or an offer or accepting a deliverable. However, a single phase of the dialogue may require the execution of several activities, such as, for example, repeated exchanges of communication acts (requests, counter-offers, withdrawal) during the negotiation phase. Because activities can cross several customer-supplier relationships, they are usually accomplished by a team of people that varies in size between two and ten. (If more than ten people need to be involved, we may be looking at a subprocess.) It also follows from this definition that activities usually include subordinate dialogues and the management of related commitments. For this reason, activities should always directly add value to the overall business process.

Activities are made up of *tasks.* A task is the largest unit of procedural work that can be executed within the same customer-supplier relationship without the need to initiate a subordinate customer-supplier dialogue. Consequently, tasks can be designed as deterministic, repeatable procedures. As such, they are the basic procedural building blocks of the business process with definable resource requirements, duration, and performance characteristics. Because of this, a task is considered the smallest fully measurable work unit within a business process.[2] Many tasks can be performed by an individual, or in some cases by a small squad of two to five people.

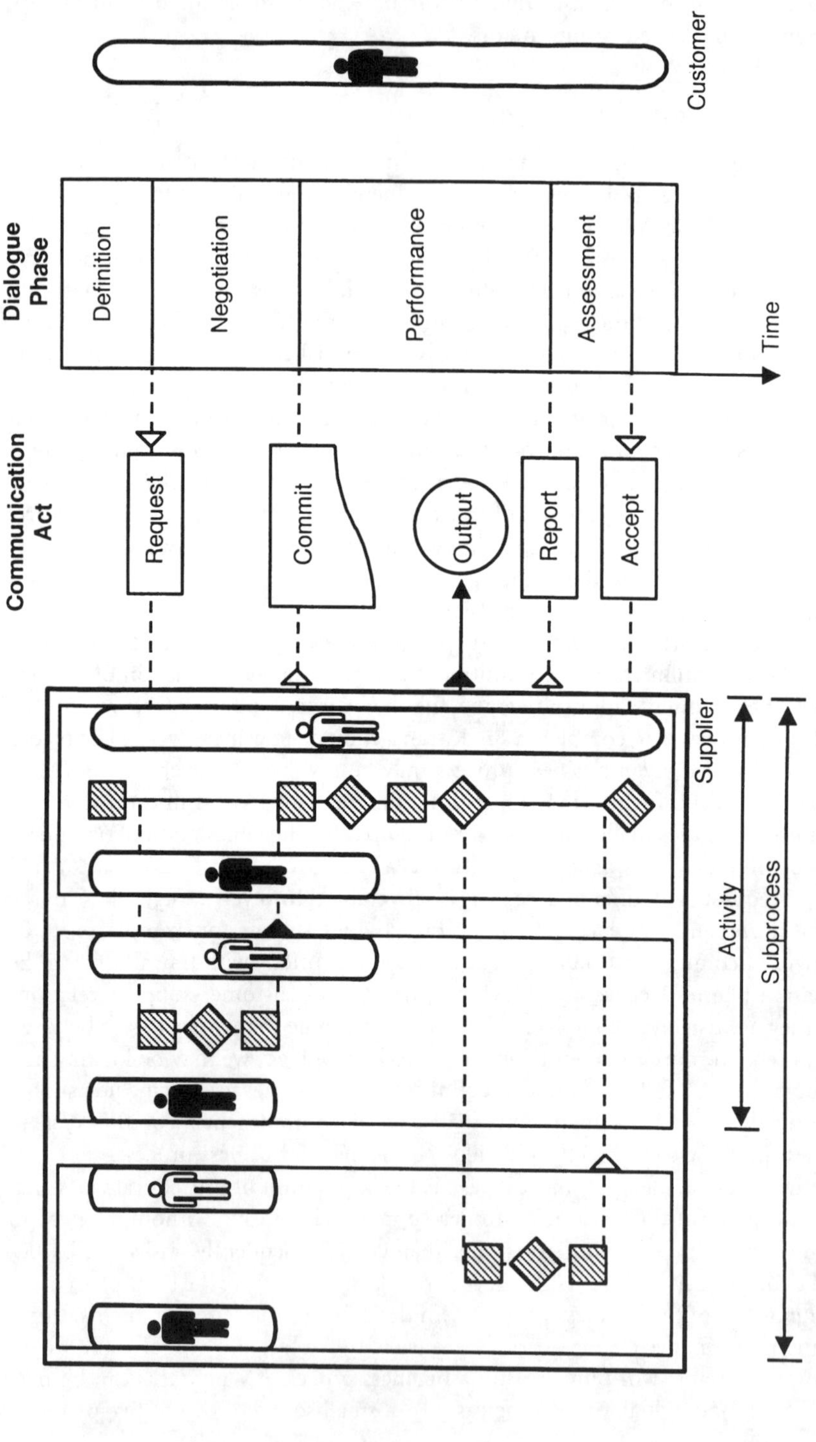

Figure 16.2. Workflow and Dialogue Alignment

Task-level planning is extremely important in software development. Meeting the commitment to deliver a major, high-quality software system requires the meticulous definition of literally hundreds of commitments by individual software programmers and teams. To ensure these commitments are met, programmers must prepare work plans that include the description and schedule for such tasks as coding, unit testing, and field testing, among others. Personal process management tools, such as the Personal Software Process (PSP)[3] and the Team Software Process (TSP) are now finding growing use in the software industry.

In many cases, the complete design of a business process does not require decomposition below the task level. In certain industrial settings, such as in manufacturing production processes, it may be necessary to further decompose tasks into a sequence of logically associated *work steps.* The work step is defined as the smallest definable and measurable unit of any executable work process; one or more steps are normally performed by a single individual. Based on their function, there are six basic types of work steps:

1. Operation
2. Transportation
3. Delay
4. Inspection
5. Storage
6. Rework

This set of work steps and the standard industrial flowcharting symbols used to represent them are shown in Figure 16.3. Also shown is a modified terminology and set of symbols more consistent with most accepted process modeling methods and the notation used in this book.

Table 16.1 shows the design template for a single task made up of several work steps. During task design, each work step must be clearly defined by name, description, and type; the business object of the work step; duration; value added; action performed; and information or data required and/or created. Such a disciplined approach to task design will ensure that the workflow being constructed will be continuous and comprised primarily of value-adding work. In professional services, it is often necessary to present highly complex business data in the form of a report. The use of the design template to describe the task of assembling, drafting, and completing such a business report is shown in Table 16.2. The task is performed by three people: a consultant, a data analyst, and an administrative assistant.

16.2 Design for Accountability

In the holistic process model, commitments result from negotiations conducted as part of a disciplined dialogue between empowered parties who play the roles of customer and supplier. Empowerment includes the authority and means to make decisions and communicate them during the negotiation phase. The most important

Standard		Business Process		Description
Step	Symbol	Step	Symbol	
Operation	○	Process	□	Value-adding action; moves workflow forward
Transportation	⇨	Move/transport	⇨	Action that moves or transports object(s) or people
Inspection	□	Inspect	◇	Inspection, check, review or approval
Storage	▽	Wait: Store/Retrieve	▽	Scheduled delay of object(s)
Delay	D	Wait: Delay	D	Unscheduled delay
Rework	Ⓡ	Rework	[R]	Unnecessarily repeated action

Figure 16.3. Six Basic Work Steps

communication is the commitment to produce the process output that will meet the agreed-upon conditions of satisfaction. The means are described in the task-level work plan linking process capabilities to commitment.

The commitment in place becomes the current performance goal of the process, and the supplier assumes accountability for achieving it. Using Scherr's original formulation,[4] accountability is the supplier's commitment to fulfill the customer's conditions of satisfaction. Accountability is outcome oriented and personal; the supplier is accountable for the committed outcome to the customer. This also means that the supplier becomes the accountable owner of the committed outcome and of the wherewithal needed to produce it. This is the accountability-based definition of process ownership. Accountability is a state. It means ownership for the duration of a single business cycle; therefore, the two are synonymous terms.

Process ownership is a management concept. It underlies Haeckel's argument that the sense-and-respond organization can only be managed through personal accountability.[5] He defines accountability simply as "who owes what to whom." The concept of identifying accountability with a committed outcome is also the basis for designing or, more appropriately, deploying accountability throughout the business process. The original customer of the business process (the primary customer in the customer-supplier value chain) has no accountability as related to process outcome, except for certain obligations assumed as part of the conditions of satisfaction (e.g., pay the invoice rendered by the supplier).

This approach to defining and implementing the concept of accountability will also help eliminate the all too frequent organizational behavior of offering the "process" — whatever that may mean — as the excuse for failures or lack of fulfilling

Table 16.1. Task Design Template

Task name:
ID:
Date:

		Work Step										
Supplier Source	*Input (Object)*	*Name/ Description*	*Type*	*Processing Person*	*Information Required*	*Actions*	*Tools Required*	*Database*	*Duration*	*Output*	*Value*	*Customer/ Recipient*

16.2. Task Design Template for Producing a Consulting Report

Task name: Produce Consulting Report
ID: 33.6/99
Date: August 15, 1999

Supplier Source	Input (Object)	Work Step: Name/ Description	Work Step: Type	Work Step: Processing Person	Work Step: Information Required	Work Step: Actions	Work Step: Tools Required	Work Step: Database	Work Step: Duration	Output	Value	Customer/ Recipient
				Consultant		reviews						
	final draft	of Consulting Report			for content							
				Data analyst		reviews						
	final draft	of Consulting Report			for logic and data quality	and creates				Final Consulting Report		
				Consultant		signs off and releases						
		Final Consulting Report										to assistant
				Assistant		prints and copies				Final Consulting Report		
				Assistant		releases				Final Consulting Report		to mailroom for mailing

Table 16.3. Commitment and Accountability Template

Dialogue Item	*Supplier (Name)*	*Customer (Name)*
Offer		
Request		
Conditions of satisfaction		
Process output		
Accountability for		
Accountability to		
Report completion		
Accept		

commitments. Individuals or groups can no longer escape responsibility by finger-pointing or simply blaming the computer, the system, or the process.

Along the customer-supplier chain, each empowered customer-supplier pair generates further accountabilities; these are identified with the supplier side of each relationship and are always linked to the particular outcome committed in that dialogue (Table 16.3). This cascading of accountabilities, however, does not diminish in any way the original accountability of the primary supplier (e.g., the business process owner), unless it is formally transferred, through renegotiation, to one or more subordinate suppliers in the value chain. In the end, the chain of suppliers in the business process will become a *chain of accountabilities* whose owners are usually the members of the permanent business process management team. (They are also part of the accountability network, suggested by Haeckel as the management mechanism of the sense-and-respond enterprise.)

In small organizations or small business processes, there may only be one person who assumes all accountability. This individual, imaginatively called the "boss," makes offers to several customers, negotiates conditions of satisfaction, and then oversees the making and delivery of products or services. In this case, each customer-supplier chain consists of only one dialogue and, most likely, one set of procedural activities under the direct control of the boss.

Once accountabilities are clearly identified, they can be incorporated into business process management and the organization as a whole, as assignments of process ownership. The organizational implications of this will be discussed in Chapters 17 and 18.

16.3 Culture of Accountability

The culture of any human organization can be described as the body of values, beliefs, behaviors, and practices characteristic of the collective membership of that organization and transmitted from one generation to the next. The culture of adaptable business processes and the process-centered enterprise is that of accountability. Its fundamental premise is that performance depends on the ability to develop accountable people who create mutual commitment and action.

What is the difference between a group of people who sincerely want to do a good job and a high-performance process team that can be relied on to deliver the result? The culture of accountability answers this in a way that leaves us with a new appreciation for what it means to be accountable. The cultural framework of accountability is based on four key concepts:

1. Communication is the essential prerequisite for action. It comprises specific linguistic forms and tools that directly define day-to-day interactions between people. A particular form of communication establishes commitment as the basis of a business relationship; it generates momentum and action, rather than just talking about action. It is this language that is used in the customer-supplier dialogue of the business process.
2. Relationship is the basis of shared success. It engenders cooperation and initiative. A particular form of relationship is the process team, comprised of empowered individuals along a customer-supplier value chain who, through teamwork, can extend their ability to achieve results by working through and with others.
3. Accountability means being committed to an agreed-upon outcome and accepting the personal consequences of having made a commitment. It is predicated on empowerment and maximum freedom of action. Accountability is a state, defined by the body of its related commitments at any given time.
4. Discontinuity is the critical event that most tests accountability and leadership and most impacts satisfactory outcomes.[6] Most examples of discontinuity are instances of rapid and unpredictable change. A special type of discontinuity is known as breakdown: the unexpected failure of the process or any of its components, usually presenting an unprecedented challenge to the process owner and the process management team.

The human carriers of the accountability culture, then, are accountable leaders and empowered teams. In this culture, accountable leadership means ownership; teamwork is characterized as the effective coordination of accountabilities. The human competencies required for success in the culture of accountability include:

- Taking the initiative for action (supporting skills: assessment and decision-making)

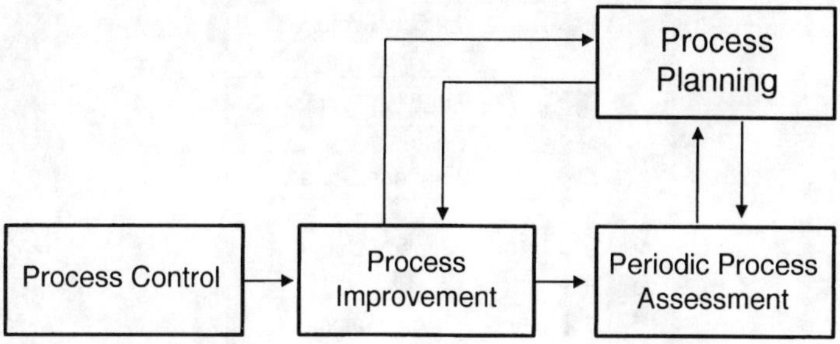

Figure 16.4. Operational Business Process Management

- Working through language (supporting skills: use of the language of commitments and mastery of the communication acts)
- Building relationships (supporting skills: developing trust, the ability to decline requests or offers, and personal involvement)

While cultures tend to evolve somewhat spontaneously over time, they can be strengthened and supported by education, training, and a purposefully planned system of organizational learning.

16.4 Continuous Improvement and Planning

Continuous improvement *per se* is not part of adaptability, but rather a pre-condition for it. Continuous improvement in a business process is defined as a series of incremental process improvements reinforced by process planning. It ensures responsiveness to more radical and unpredictable change by creating an existing base of process effectiveness and efficiency. This is achieved through the continuing elimination of costly process waste represented by unnecessary and non-value-adding work activities and tasks; therefore, continuous improvement must be designed into the operational management of the business process.

Both process improvement and process planning assume effective process control to be already in place (Figure 16.4). As we have seen earlier, process control is designed to maintain the stability of procedural workflow during execution or restore it to stability by eliminating the causes of sporadic variation. Restoring a process to its intended operating state is also known as *correction*. However, we have shown that the concept of control can be extended to the commitment dimension of the process as well, through the insertion of feedback learning loops into the communication flow between customer and supplier (Figure 14.1). Continuous process improvement and planning, as ongoing operational activities, are also prerequisites for process adaptability.

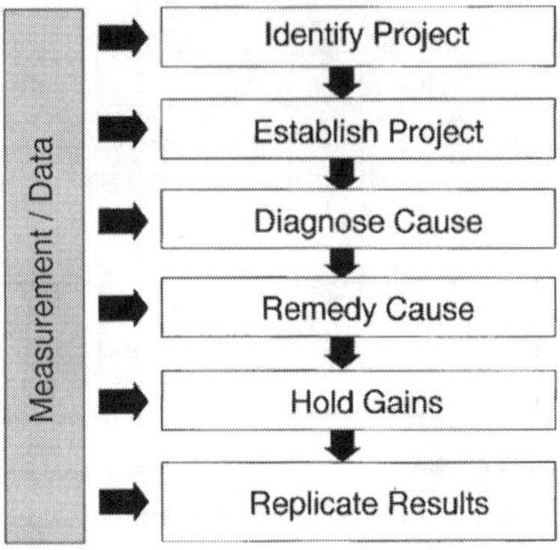

Figure 16.5. Process Improvement

16.4.1 Process Improvement

Process improvement becomes necessary when process control uncovers chronic deficiencies it cannot handle by simple corrective action. A chronic problem is one that persists over time and is not solvable with the methods, tools, and resources of control. Process improvement is always aimed at what exists; it will either (a) eliminate it, or (b) change it in function or structure (but not both) in order to attain beneficial process change.[7]

Process improvement is defined as the permanent removal of the causes of chronic deficiencies. It is usually implemented as a six-step sequence of actions by a process improvement team in the framework of an improvement project (Figure 16.5). The team analyzes a chronic problem within the business process to determine the problem cause and to implement the remedy, which will eliminate or mitigate that cause. Process improvement projects are almost always initiated as *ad hoc* events by the process owner. A subordinate customer/supplier dialog is usually established to secure the special commitments of the project. While most of the tools and techniques of process improvement are applicable to the procedural workflow, the improvement concept can also be extended (just as control) to the commitment dimension of the process as well.

16.4.2 Process Planning

Process planning is aimed at what does not yet exist; it leads to incremental changes that are needed or intended. Process planning is defined as the design of new process capabilities in terms of structure or resources.

16.4.2.1 Reverse Planning

In line with the requirement for maximum process adaptability, the type of planning used is known as reverse planning, first suggested by Ackoff as "planning backward."[8]

It is the planning approach increasingly used in the face of discontinuity. Traditional planning and most strategic planning methods are based on going from where one is now to where one will want to be at some point in the future; it is working from the present to the (presumably) predictable future. Unfortunately, in today's environment, there is just too much unpredictable change, occurring too quickly, to be able to forecast reliably what will happen in the future. For this reason, where we now think we will want to be at some point in the future is most likely very different from where we will want to be at that same future point in time.

Reverse planning, on the other hand, works backward from where one wants to be now to where one is now; it is working entirely with information available in the present. Contingency planning is a form of reverse planning, which uses reverse planning to identify situations where one wants to, or could, be now. These situations will then represent a range of possibilities, where a contingency plan can be developed for each. This approach is taken when process capabilities must be identified in advance for rapid execution in support of new or changed commitments.

16.4.2.2 The Year 2000 Problem

Alternatively dubbed "time bomb 2000," the "millennium bug," or the "challenge of the century," the real and imagined problems related to entering the year 2000 have lent themselves to much speculation and even hype. Instead of being terrified by some of the doomsday scenarios promoted by the press, pundits, and some authors, management in most organizations has arrived at a reasoned conclusion about what might happen and has decided on appropriate measures to deal with the issue. One of these measures is contingency planning.

What is the Year 2000 (Y2K) problem?[9] Incorrect date-dependent decisions could be made by automatic devices (mainframe computers, personal computers, microprocessors, controllers, and the like) in the year 2000 and beyond, resulting in outcomes that range from the simply inappropriate to the outright catastrophic. Two fundamental causes of the Y2K problem are that (1) automatic devices may not recognize the year 2000, and (2) they may not recognize the year 2000 to be a leap year.

In plain English, when a computer does not know that it is in the year 2000, it may arrive at incorrect decisions that depend on a specific date within that year (or beyond). Also, if the computer does not know that the year 2000 is a leap year, it will not recognize that the 29th of February exists and, consequently, all its decisions related to that date will be invalid or incorrect.

The Gartner Group has estimated that the total cost of preparing for and dealing with the Y2K problem worldwide will be between $300 billion and $600 billion. Other estimates have ranged between $200 billion and $850 billion worldwide.[10] Whatever the final amount will be, it is certain that preparing for and dealing with the Y2K problem will have cost money — and to some, unaffordably large sums of it.

Y2K is not just an information technology problem to be solved by information technologists. While it is rooted in computer technology and, in particular, date-dependent software applications, its consequences may be global, affecting institutions, organizations, and individuals alike. *Y2K is a pervasive business process problem.*

Moving from 1999 into the year 2000 is like crossing a minefield which one knows has been there but which one also hopes has, in the meantime, been cleared. However, since no one knows how many mines have been laid and how effective the clearing operations have been, it is inevitable that some mines will be detonated and some damage will be caused when the crossing of the minefield does begin on January 1, 2000. This is where contingency planning comes handy.

Contingency planning does not intend to offer a perfect solution because such a solution would mean the detection, correction, or removal of all date-dependent decisions in existing software that could result in undesired outcomes. Pieces of the solution have, by and large, been identified and their implementations are underway. Contingency planning needs to deal with the fact that in problems this massive, some of the required solutions will not be completed in time.

Our focus, therefore, is on January 1, 2000, and beyond. Once the minefields have been crossed, the damage, if any, must be dealt with promptly. Focus must be on preparations to minimize damage and deal with damage if and when it occurs, as well as planning for recovery. This is the classic case of planning for recovery from any number of possible failure situations. It is a special form of the reverse planning approach described above.

As the exact consequences of incorrect date-dependent decisions in the computer systems of banks, utilities, government, and other organizations (and in the personal computers and various microprocessors belonging to individuals) are unpredictable, most contingency planning for Y2K problems is taking the form of *scenario planning*.

Scenario planning specifies a range of possibilities for an anticipated event, whose magnitude or exact nature cannot be predicted. It then identifies a set of discrete, likely events within that range known as scenarios. Using reverse planning techniques, a contingency plan is developed for each scenario by identifying the process capabilities required to execute it.

16.5 Summary

The holistic process model provides the basic design structure for process adaptability. Once this structure is defined, the design for adaptability is carried out in three steps: (1) defining the backbone network of commitments, (2) establishing the reconfigurability of the network and the work activities it drives, and (3) ensuring robustness through the deployment of accountabilities throughout the process.

In the holistic process model, commitments result from negotiations conducted as part of a disciplined dialogue between empowered parties who play the roles of customer and supplier. The commitment in place becomes the current performance goal of the process and the supplier assumes accountability for achieving it.

The design for accountability is essentially the integration of the commitment network (a structural concept) with process ownership (a management concept). Once accountabilities are clearly identified, they can be incorporated into business process management and the organization as a whole, as assignments of process ownership.

Process adaptability is further enhanced by a culture of accountability — developed and nurtured by people who create mutual commitment and action. The cultural framework of accountability is based on four key concepts: communication, relationship, accountability, and discontinuity (the change that tests accountability the most).

Continuous process improvement and planning, as ongoing operational activities, provide the prerequisite current base for process adaptability. Both process improvement and process planning assume effective process control to be already in place. Process improvement is aimed at what already exists. It becomes necessary when process control uncovers chronic deficiencies it cannot handle by simple corrective action. Process planning is aimed at what does not yet exist; it leads to changes that are needed or intended. An interesting case in point for contingency planning, a form of process planning, is the ongoing effort to anticipate and minimize problems associated with the arrival of the year 2000, the so-called Y2K problem.

17 Launching the Process

The transfer of the reengineered process from plan into reality is accomplished in three steps:

1. Planning for implementation problems
2. Planning the actions of implementation
3. Deploying and executing those actions within authorized resource levels and on schedule

These steps are carried out in an integrated fashion, across process structure, information technology, and human organization.

After a new process design is completed, the process management team moves to implementation. Implementing a new process is usually the most failure-prone phase of process management because of the organization's inertia and even resistance to the change represented by the new process design. Thus, the challenge during implementation is not just working out the structural, technical, and resource-related aspects of moving to an entirely new way of doing things, however complex they may be, but managing the human and organizational dimension of change, as well. Assertive and competent leadership by the process owner and a sound implementation plan are essential to success.

17.1 Process Implementation Planning

Process implementation planning as used in this book refers to the activities of planning the introduction, launch, and operational implementation of the business process — that is, planning how to make it work. The end result is the *process*

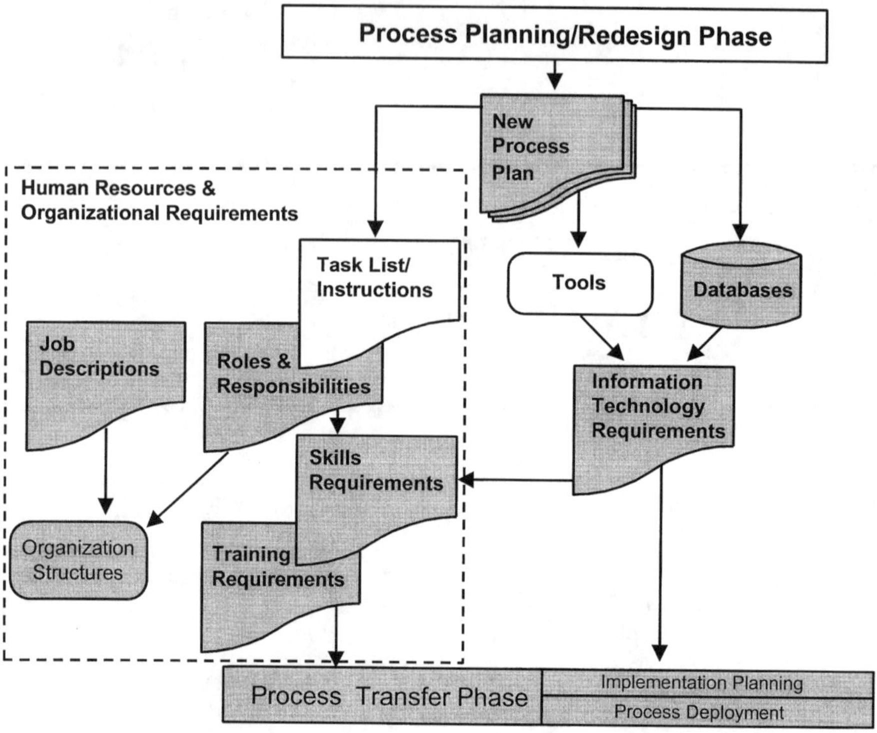

Figure 17.1. Process Redesign and Implementation

implementation plan. Process deployment refers to the actual implementation of the activities planned, managed with the help of an implementation project plan and budget. Implementation planning and process deployment together make up the distinct process management phase known as process transfer[1] (or process launch), so named because it takes the results of process planning and develops the methods, means, and resources to transfer what is essentially a "process blueprint" into operational reality.

The process transfer phase follows the preceding phase of process planning — discussed in the previous chapters. Inputs to implementation planning are the results of process planning: the process design documented in the process plan (see Table 15.1), accompanied by human resources and organizational requirements and information technology requirements (Figure 17.1).

Business process redesign can be carried out as one of six process planning alternatives characterized by their design purpose (Table 17.1). The seventh alternative — essentially the "do nothing" case, designated as "current state" in the table — represents no change and, therefore, no redesign effort. Of the remaining six, the first five can be attained by merely improving or redesigning the workflow part of the process — the traditional process reengineering approach. The last redesign

Table 17.1. Process Planning Alternatives and their Results

Process Planning Alternative	*Design Purpose*	*Process Result*
1	Current state	*Status quo*; no change
2	Span of control	Streamlined management structure
3	Logistics focus	Efficiency
4	Quality focus	Effectiveness
5	Process focus	Effectiveness and efficiency
6	Variance controlled	Process in control
7	Commitment controlled	Effective, efficient, and adaptable

alternative represents the holistic approach and is the only one that materially contributes to process adaptability. In the discussion that follows, we will assume that implementation planning deals with the results of this type of process design.

17.1.1 Integrated Implementation Planning

In its description of the customer-supplier value chain and the associated relationships, the process plan includes a number of paired role definitions (e.g., customer and supplier) related to specific positions or individuals involved in commitment management. The customer-supplier dialogue defines the roles and responsibilities associated with commitment management. The human resources and organizational requirements, on the other hand, contain all the individual roles and responsibilities derived from the workflow dimension for everyone involved in the business process.

The sum total of these roles and responsibilities, both at the individual and team level, specify what people are expected to do and how, day after day on an ongoing basis, once the business process becomes operational. The process plan also identifies process resources that will be required to carry out all operational roles and responsibilities — and to perform all the work activities and communications called for by the process design.

However, roles and responsibilities must be prepared for; the preparations (such as training, job change, and the like) are planned and carried out in the process transfer phase. The execution of these preparations is part of the effort needed to launch the new process. Therefore, above and beyond the resource level required for the future operation of the redesigned business process, implementation planning establishes additional resources that will be needed for launching the business process into operation. This resource is required only for a finite period of time and, because of this, is also known as the implementation "resource bubble" (Figure 17.2).

The takeoff of an airplane is a good analogy for this type of resource requirement. The plane is the embodiment of the process of flying; it is designed to perform this with the support of certain resources while in flight (cockpit crew and flight

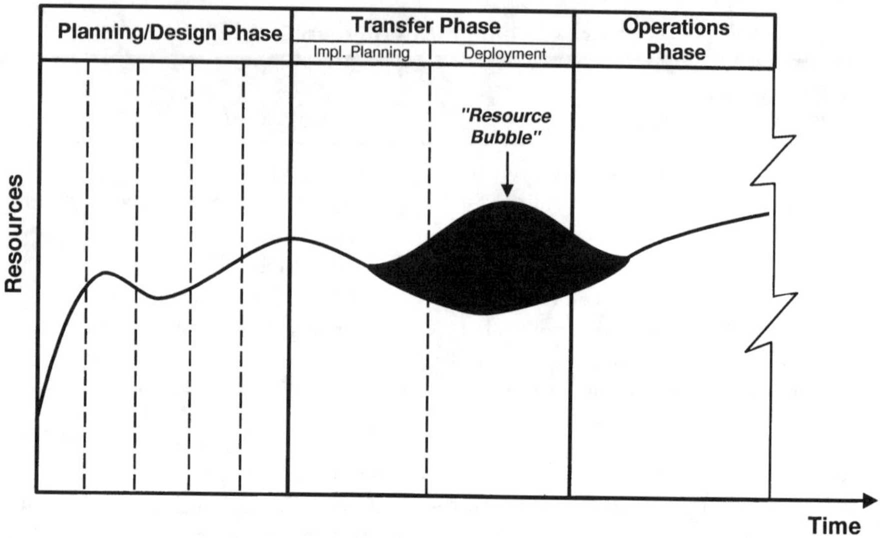

Figure 17.2. Process Reengineering Resource Distribution

assistants, fuel, supplies, and so on). However, getting airborne requires additional resources: gate personnel and the ground crew, air traffic controllers, runway, communications, higher rate of fuel consumption, and so on.

17.1.2 The Three Steps of Implementation Planning

In essence, process implementation planning consists of three distinct steps (Figure 17.3):

1. *Step One: Planning for implementation problems:* (a) Identifying and then planning the removal of existing constraints (inhibitors, barriers, resistance), and (b) identifying new support requirements (often based on gap analysis) and the enablers that satisfy them.
2. *Step Two: Planning for action:* Developing action plans to deal with the problems identified in the first step. An important component of this step is communication planning — creating a strategy for communicating progress within the business process, within the enterprise, and to outside parties, such as external suppliers and communities.
3. *Step Three: Process deployment:* Mobilizing and deploying the resources needed, and implementing the action plans identified in Step Two.

On occasion, a fourth step may be added:

4. *Step Four: Planning of several implementation alternatives and the selection of the preferred one.*

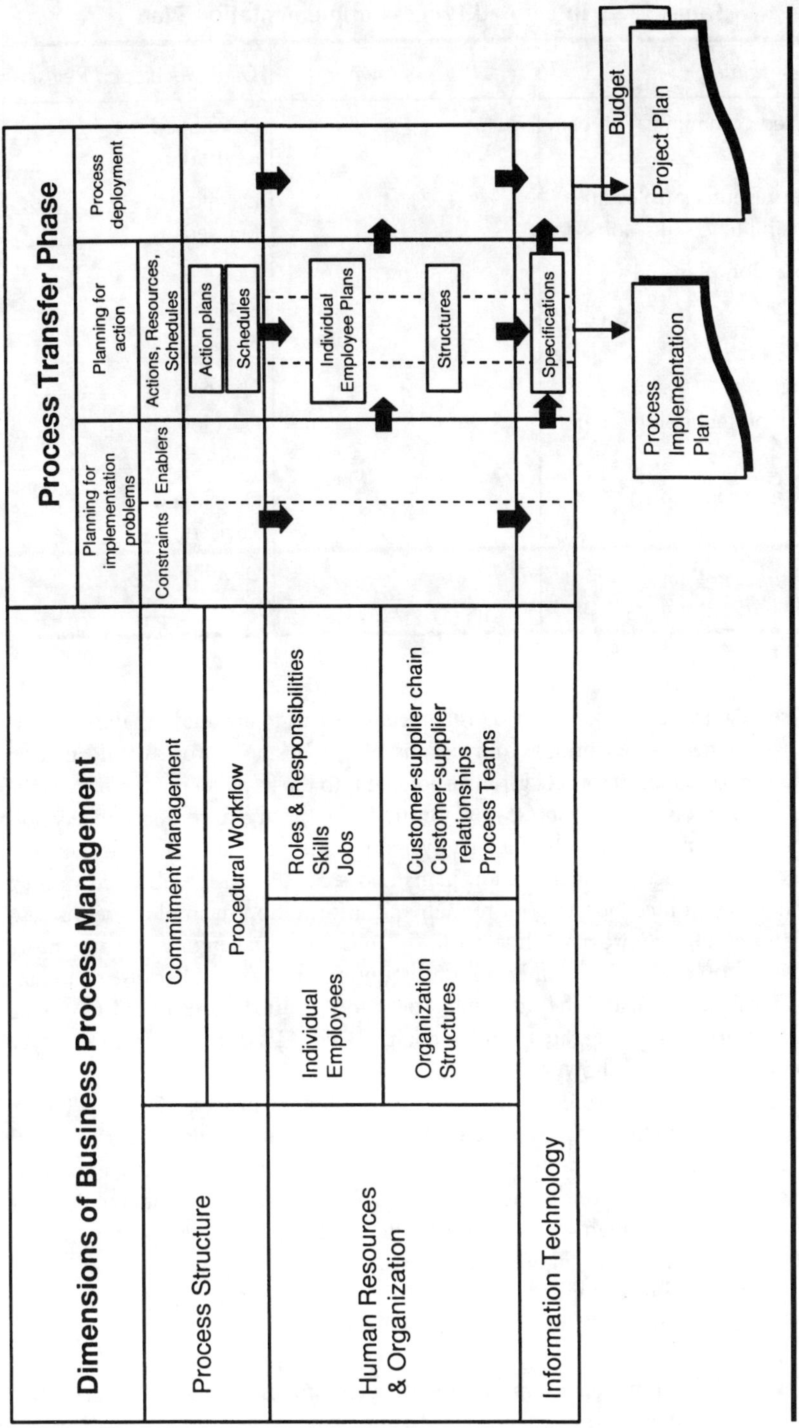

Figure 17.3. The Process Transfer Phase

Table 17.2. Integrated Process Implementation Plan

Process Name	*Process Owner*	*Date Developed/Revised*
Integrated communication plan		Developed: ________ Revised: ________
Human resources (roles and responsibilities, skills, jobs)		Developed: ________ Revised: ________
Organization plan		Developed: ________ Revised: ________
Training plan		Developed: ________ Revised: ________
Information technology plan		Developed: ________ Revised: ________
Integrated project plan		Developed: ________ Revised: ________
Budget		Developed: ________ Revised: ________

These steps must be accomplished in an integrated way, to include both the commitment management and workflow dimensions of the business process. Information technology and human resources get incorporated into the implementation plan as the key process assets. Figure 17.3 shows how constraints in the organizational and technology areas are identified in Step One. To overcome these and also to implement key enablers of the process launch, action plans and schedules are established in Step Two. These again incorporate both human and information technology resources.

This integrated approach, while more complex and challenging, ensures more rapid overall progress and a minimum of disruption and rework because of inconsistencies, mismatches, and the subsequent need to retrofit. The result of this overall planning effort is the integrated implementation plan. Its content, representing a series of deliverables, is shown in Table 17.2.

Implementation of the reengineered order-fulfillment process (OFP) in Union Camp's Fine Paper Division was planned by using this integrated approach (Figure 17.4). The information technology solution was a Baan enterprise system, as parallel implementation planning under the process design umbrella was considered the only possible alternative. A high-level version of the resulting integrated process implementation plan is shown in Figure 17.5. Integrated implementation planning is also consistent with the holistic approach to reengineering; most importantly, however, it ensures the adaptability of the newly implemented business process from the very beginning.

Below, a discussion of the implementation-planning phase examines each of its three steps in sequence, and an integrated view of the three process components

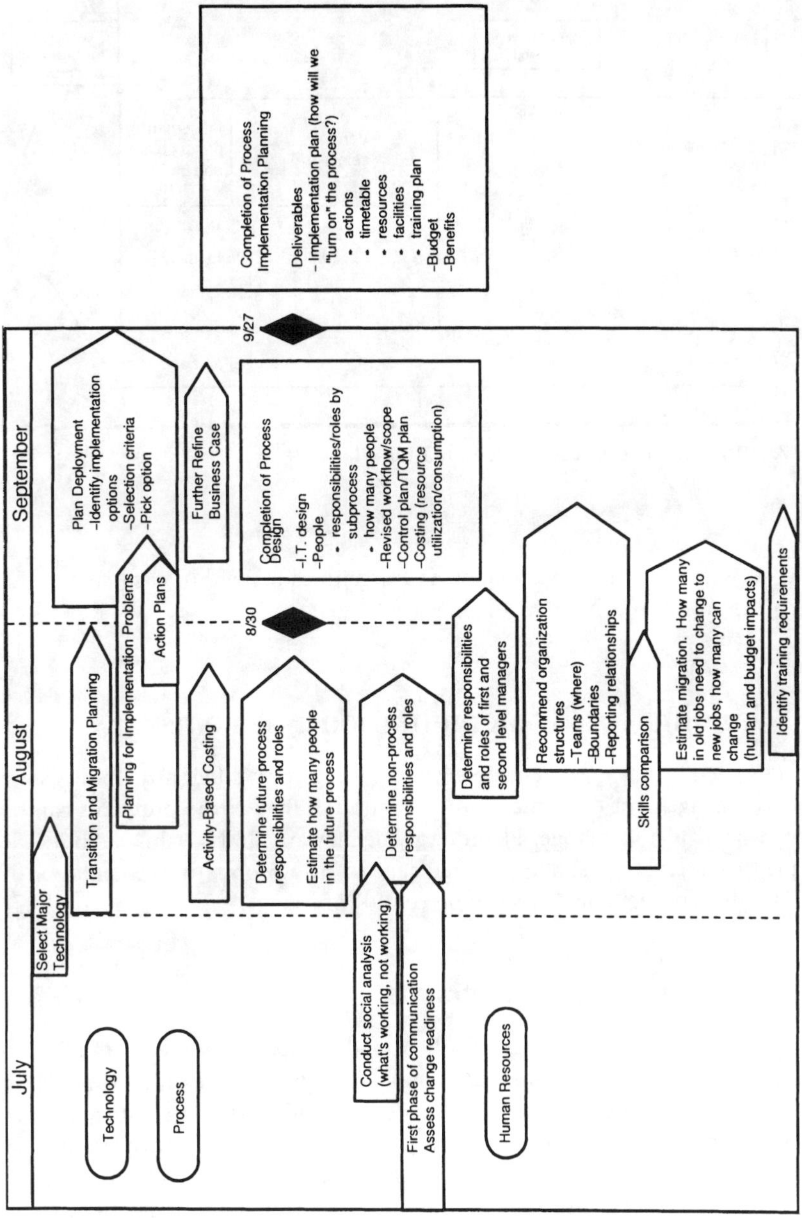

Figure 17.4. Integrated Implementation Planning

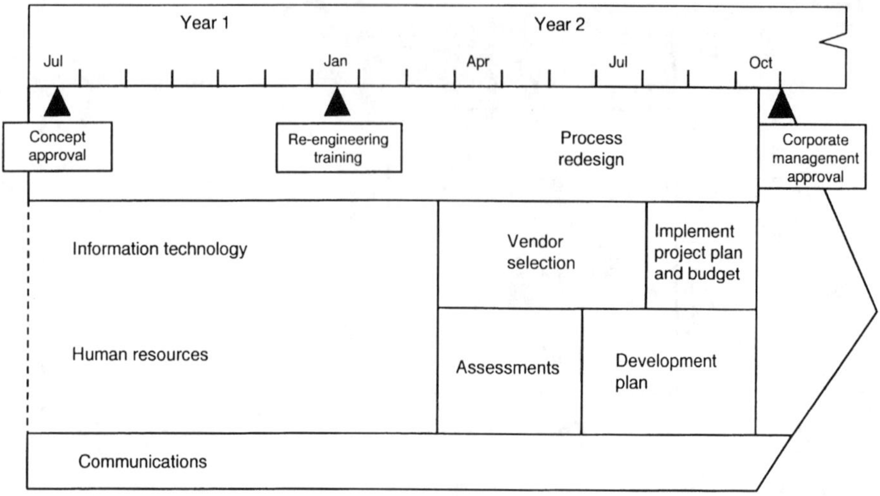

Figure 17.5. Integrated Process Implementation Plan

(business process structure, technology, and organization) within each step and a real-life planning approach are highlighted.

17.2 Planning for Implementation Problems

A newly designed business process means doing things very differently. In this step, the process management team makes the assumption that all the problems can be anticipated prior to the change. Planning guidelines are always valuable, but it is important to remember that the real challenge of new process implementation is dealing with the consequences (real or perceived) of change.

17.2.1 Company-Wide Constraints

In every implementation planning effort, overall company-wide constraints and inhibitors should receive attention first. These are usually represented by existing policies, standards, accounting rules and guidelines, and legal or regulatory restrictions. In the course of planning, changes to policies are usually incorporated in procedural changes. Many times, policy changes require higher levels of approval, and it is here that the process stakeholders or the transformation council can expedite resolution. Certain standards, rules, or guidelines may be removed or amended; others, such as legislation or regulation, which may require long-term lobbying efforts, cannot be changed and will have to become assumptions and conditions around which implementation can proceed.

17.2.2 Process-Level Impediments

The types of impediments encountered in implementing the business process structure fall in two groups: those related to commitment management and those pertaining to the workflow.

17.2.2.1 Problems with Commitment Management

Commitment management is new to most organizations. For this reason, implementation planning must carefully attend to the constraints that may be encountered. Existing cultural or behavioral patterns and traditional, job-oriented work practices may slow down the launch of a disciplined and effective customer-supplier dialogue. The solutions to be planned here include, at a minimum:

- Clear definition of the roles of customer and supplier as peers, regardless of their position or organizational status, along the entire value chain; these roles also comprise the concept and definition of accountability
- Clear and concise definition and practice of the customer-supplier communication protocol: the form, content, and meaning of the seven possible communication acts within the customer-supplier dialogue, and the rules for their effective use
- Clear and complete definition of the customer-supplier dialogue: the specific sequence of all the communication acts and related supporting procedural activities required to complete a single business cycle (usually representing a subprocess)

Each of these solutions is supported by appropriate orientation and training.

17.2.2.2 Problems with Workflow

Also, technological factors may impede the implementation of reliable, continuous, and rapid communications along the value chain and within individual customer-supplier relationships. An outdated network architecture and related facilities, inadequate bandwidth and transmission capacity, limited connectivity, poorly organized network management, and the like may present a serious implementation problem. The remedies here include upgrading the network architecture, effective network management and controls, and state-of-the-art performance monitoring equipment and software.

Implementation of the new process workflow may run into different obstacles:

- Inconsistent terminology and definitions to distinguish the components of process flow, which includes the flow of work, materials, and information
- Unclear definition or lack of understanding of the various units of work (activity, task, work step)
- Missing or incomplete measurements and controls

17.2.3 Cultural Resistance

In the area of human resources and organization, implementation of a new process design that includes new roles and responsibilities is often considered an unwelcome change to the norm, at best, and a threat to status and job security at worst. Because of this, people will resist it in various overt or covert ways, a phenomenon Juran calls cultural resistance.[2] Change is resisted on account of its intended outcome and social impact.

Of the two, the latter poses the bigger challenge because social impact often is not only real, but also personal. For example, the introduction of process ownership and the related accountability goes hand in hand with empowerment: the authority and means to make decisions and commitments. This authority and associated means represent newly acquired power, which must come from somewhere within the enterprise. Because power is a finite commodity in every organization, it must be transferred to the process owner from someone else, usually the functional management. This is the real threat to the traditional management hierarchy.

Thus, at any time, a perfectly workable new idea can be rejected by those in charge because it is seen as threatening. Reasons can be stated or real. Typically, stated reasons (e.g., it won't work here; it has already been tried without success; it's too costly) hide the real ones (e.g., my job will change; my resources will be reduced; my authority will be diminished). In planning to deal with cultural resistance, three key actions will help.

17.2.3.1 Ownership and Accountability

Clear ownership and accountability along the value chain must be assigned. This will make many of those affected by the change also own it and the solutions to it. Dealing with at least part of the resistance becomes the responsibility of the process management infrastructure. Ownership will force the resistance to be resolved.

Cultural resistance by the established management hierarchy may have a further unwelcome consequence. The perception of slow progress (or none) in implementing needed changes will drive dynamic and progressive managers and professionals out of the organization. To prevent this from happening, ownership and accountability should primarily be assigned to mobile, forward-looking people who are more willing to take the associated risks.

17.2.3.2 Surprise-Free Environment

A surprise-free environment can be maintained through frequent and effective communications. Developing a comprehensive communication strategy and detailed plan is one prerequisite to success in dealing with resistance. One of the key contributors to the general acceptance and successful implementation of the newly reengineered order-fulfillment process at Union Camp was a thoughtfully planned communication strategy, consistently executed throughout the design and implementation planning phases. The overall strategy divided communications into two major categories: those within the process and those external to it (within and outside the company). Communications within the order-fulfillment process followed a basic plan that established a general pattern and was updated on a regular basis to match the progress of the reengineering effort (Table 17.3).

As a result of this strategy, a great deal of the expected resistance to reengineering never materialized because senior management made sure that everyone in the organization knew exactly what was happening and why the proposed implementation was so critical to the continuing success of the company. The subject was kept visible to all employees through bulletin board announcements, newsletters, and group briefings. While expected changes and their anticipated results were continuously being reported to the organization at large, there was an equal amount of attention given to feedback about concerns regarding the reengineered process and suggestions for its more effective implementation.

17.2.3.3 Minimizing Resistance to Change

A set of rules of the road as guidelines for preventing or minimizing resistance should be adopted from the outset, including:

- *Promote participation.* Involve as many impacted people as possible, which is normally done through surveys to get employees' ideas and opinions, small group discussion with process team members, and one-on-one reviews by the process owner with key stakeholders.
- *Provide time for the change.* Allow for reasonable lead times before implementation; allow ample time for training and orientation.
- *Work with the informal leadership.* Identify and enlist the help of "opinion makers" and "political leaders" in key change-affected areas. Ensure that they are involved in critical implementation planning and discussion events. Better still, make the new process design coincide with the very same ideas they may have been seeking to promote.
- *Reverse positions.* During implementation planning, members of the planning team role-play the part of those affected by the change. Their concerns and reactions will be easier to identify and understand from this perspective.
- *Treat people with respect and dignity.*

Beyond cultural resistance, the two most frequently encountered constraints in human resources are (1) the wrong mix of knowledge and skills for performing newly introduced work steps or tasks, and (2) misunderstanding and even confusion about new roles and responsibilities — both at the individual and team level. The first constraint can usually be overcome by appropriate orientation and training, followed by coached or supervised entry into work practice. The second constraint can be dealt with by creating organizational entities (for example, process or subprocess teams or functional groups) as the collection of related roles played by groups of people.

Table 17.3. Communication Plan for the Order-Fulfillment Process of Union Camp's Fine Paper Division

Communication Events		*Audience Coding*				*Communication Method*				*Time Frame*			
Target Audiences	*Message*	*A*	*B*	*C*	*D*	*Live*	*Team Talk*	*Video*	*Other*	*1H 1996*	*3Q 1996*	*4Q 1996*	*As Needed*

Audience codes:

A = Within the order-fulfillment process

B = Major order-fulfillment interface/support (Sales, Information Services)

C = Outside the order-fulfillment process

D = Others (communities, contractors)

Roles are assigned to persons: any one of them can play several roles within the same or different processes; however, any given individual can only play one role in any one process at a given point in time. For example, the same person can first play the role of supplier to the external customer of the process — in what is known as the primary customer-supplier relationship. He or she can then turn around to play the role of customer when requesting work to be done further down in the process, in support of the commitment he or she is about to make to the primary customer.

Another example is management. All managers in a given organization play at least two roles: employee and supervisor; however, while they are in the employee role in a given process, they cannot play the role of supervisor in that same process at the same time. For instance, a manager may play the role of employee (person being appraised) in the performance evaluation process, but he or she can play the role of supervisor (person giving the appraisal) only in another instance of that same process and with another individual (playing the employee role).

The job description of an individual can be derived from all the roles that have been assigned to him or her: it is the generic list of responsibilities, the totality of things (work activities, training, communications, and so on) the employee has to do in playing those roles. A convenient instrument for generating a person's responsibility list and then his or her job description is the task design template (Table 16.1). Taking into account all the tasks this individual is involved in, the sum total of work steps he or she is expected to carry out within those tasks will yield a unique list of responsibilities which can then be converted into a generic job description. Finally, reporting relationships can also be derived from an individual's assigned roles, because a role is always stated in context and in relationship to other individuals, groups, or the organization as a whole.

17.2.4 Technology Constraints

In information technology, the constraints are usually technical in nature. The most prevalent among these, and also causing the most headache and challenge, is the existence of out-of-date or even obsolete systems. Most of these systems (also known as legacy systems) were designed and installed many years, if not decades, ago. They are based on outdated architectural concepts and are for different users and applications operating in a different organizational context; furthermore, the performance of these systems is usually unsatisfactory. But, because of the considerable investment they represent and the potentially enormous cost of changing to a state-of-the-art solution, the inertia of legacy systems can pose a very serious problem.

It is beyond the scope of this book to discuss the ways of overcoming the challenge of legacy systems. Suffice it to say that the persuasive case for technology change is always based on value assessment and future benefits to the business process, especially in terms of adaptability.

Information technology-related planning and development typically follow system development life cycle (SDLC) methodologies. Technology deployment culminates in the live operation of all aspects of the new information system: hardware, software, interfaces, data, telecommunications, and user procedures.

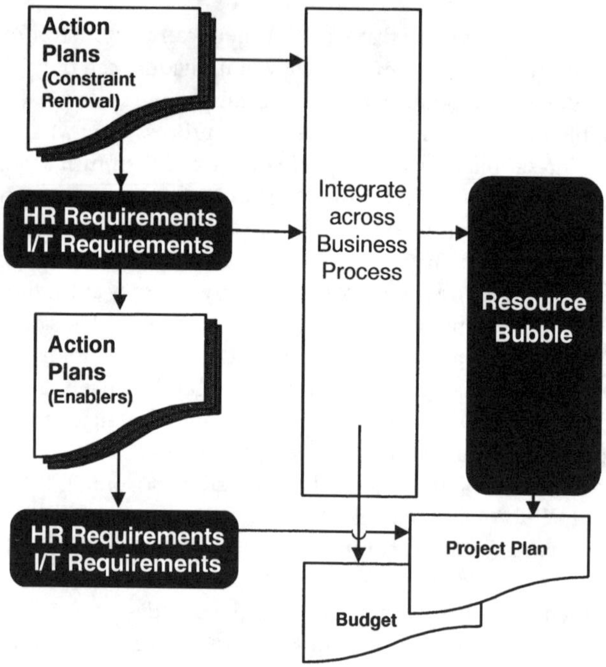

Figure 17.6. Process Implementation Planning

17.3 Planning for Action

The previous step has established opportunities for the removal of constraints. It has also identified possible enabling initiatives for bridging gaps between current capabilities and to those prescribed by the new process design. These must now be translated into action plans, specific courses of action designed to attain stated objectives with specified resources and on a specified schedule.

The action planning effort produces two sets of action plans: one for the removal of constraints, and the other to provide needed new enabling capabilities. Each of these sets also has an associated, incremental set of requirements for human resources and information technology (Figure 17.6). One important implementation action to be planned is the pilot testing of the reengineered new process. Pilot testing is essential to confirm the soundness of the reengineered process in actual practice and to reveal problems in both technology and the so-called soft areas involving human resources and organizational issues. Successful pilot testing builds support for the full-scale implementation, and it is especially critical in securing the funding needed for further progress.

The totality of resources required for the execution of all action plans (including the related human resources and information technology requirements) make up the

Table 17.4. Action Planning Template

Implementation Action		*Skills*	*Job*		*Information Technology Requirements*		
Designation	*Duration (days)*		*Description or Title*	*No. of Jobs*	*Database*	*Tools*	*System (Hardware/Software)*

One of the information technology requirements resulting from the reengineering of the order-fulfillment process at Union Camp was the radical upgrading (reengineering) of its corporate-wide network planning and management process. Dubbed the "totally reengineer Union Camp networking" process and abbreviated as TRUCN, its process team used action planning templates during the implementation planning phase. A specific example of a completed template is shown in Table 17.5.

The set of action plans with their related resources and schedules represent one implementation alternative. In many cases, this is the only one needed; however, for highly complex, overarching business processes or when several newly reengineered business processes are to be implemented, several implementation alternatives may have to be evaluated and the optimal one (or the optimal sequence) selected.

The evaluation and selection of process implementation alternatives require an optimization approach based on the measurement and comparative analysis of key performance indicators across all alternatives. These indicators may include implementation cost, scheduling, human skill and resource requirements, complexity, technical difficulty, and environmental impact, just to name a few. Other frequently used methods include value analysis supported by various cost/benefit calculations. In addition, simulation modeling may be used to project the myriad of changing interactions and dependencies during implementation.

17.4 Process Deployment

Simply stated, the individual action plans with their associated resource requirements and schedules must next be consolidated into coordinated management instruments: the integrated implementation project plan and budget. The project plan is the integration of all the implementation action plans and all related resources and schedules. In view of this, its management as a whole will essentially also represent an integrated approach. In most cases, traditional project management methods and techniques (supported by appropriately chosen software tools) are used for this purpose.[3]

Where an enterprise resource planning (ERP) solution is an integral part of the business process reengineering effort, the project planning and management function of the ERP system may be invoked. Major ERP systems, such as SAP and Baan among others, offer these capabilities as part of their system packages.

The process implementation budget contains both expense and capital items. It is always preferable to establish a consolidated process reengineering budget, as opposed to distributing the cost implications among the affected functional operating budgets. All too often, plans fall behind schedule or even fail because functional managers are usually not motivated to update their process budgets in a timely manner, especially if the proposed process changes do not directly contribute to their functional objectives.

Costing methods used in establishing the implementation budget should be consistent with those employed to plan and measure the continuing process costs once the reengineered business process has become operational. Because most process costing methods are activity or task based, it is recommended that activity-based costing also be used in building the process implementation budget, called by some the "bubble budget."

It is important to point out that all implementation activities in the project plan must contain, beyond people and technology, all the required resources: materials, energy, plant and facilities, intellectual capital, and other assets needed to ensure the successful plan implementation. While it is rare, it may happen that the deployment of a newly reengineered business process may take more time and effort than its design. This will more likely be the case where a massive upgrading of information technology, such as the implementation of an ERP system, is part of the reengineering effort.

The redesign phase for Union Camp's Fine Paper Division's order-fulfillment process lasted 14 months. During this time, a new process management infrastructure was established and the process flow completely redesigned, including the high-level information technology design. Process implementation planning took another 2 months and included the decision to implement a Baan enterprise system solution. At that point, corporate management's approval was secured to proceed. Over the next two years, the staged effort of deploying the new Baan system solution was the dominant factor in the implementation effort.

As we have stated earlier, implementation is a key and decisive phase of process reengineering. Because of this, all members of the process management infrastructure are involved in the entire effort in specific ways and on an ongoing basis, as shown in Table 17.6.

17.5 Summary

The transfer of the reengineered process from plan into reality is accomplished in three steps: (1) planning for implementation problems, (2) planning the actions of implementation, and (3) deploying and executing those actions within authorized resource levels and on schedule. These steps are carried out in an integrated fashion, across process structure, information technology, and human resources and organization. The process transfer phase follows the preceding phase of process planning (discussed in the previous chapters). Inputs to implementation planning are the results of process planning: the process design documented in the process plan, accompanied by human resources and organizational requirements and information technology requirements.

The action planning effort produces two sets of action plans: one for the removal of constraints, and the other to provide needed new enabling capabilities. Each of these sets also has an associated, incremental set of requirements for human resources and information technology. The result of this overall planning effort is the integrated implementation plan. Once the action plans are in place, they must next be consolidated into coordinated management instruments: the integrated implementation project plan and budget. Where an enterprise resource planning solution is an integral part of the business process reengineering effort, the project planning and management function of the ERP system may be invoked.

Table 17.5. Implementation and Staffing Plan

					I/T Requirements		
Action Designation	*Duration (days)*	*Skills*	*Job Description*	*No. of Jobs*	*Data-base*	*Tools*	*System*
Network plan specification	5	Project management	Network planner	3		MS project,	Windows95
		Network planning	J, T			MS PowerPoint	
		Network experience/ reference	External network consultant	1		VISIO, Auto Discovery	
Network measurement requirements	1	Network management	Network manager	8			
Network measurement tool specifications	3	Network engineering	Network engineer				
Network measurement tool selection/acquisition	20	Product selection	Reference material				
		Procurement facilitation					
Perform baseline measurement criteria	3	Network management	Network manager, T	3	Yes		
		Network engineer	Network engineer				
			Consultant	1			
Design SLA template	2	Consulting	Network manager	1			
		Customer Service	T, H				
		Network engineering	Network engineer	1			
		SLA design	First-tier team member	1			

Negotiate and publish SLAs	20	Consulting	Network manager	1	Yes	Estimating tools
		Customer service	Network planning	1		
		Network engineering	Network engineer			
		Customer decision-making		1		
First tier		Network analyst	Network analyst	1	Yes	First tier tools
Checklist	2	First tier experience	First tier (analyst)	1		
Training in tools	2	Training skills	IS trainer	1		
Access to tools	1					
Second tier	1	Network analyst	Network analyst	2	Yes	
Service standards		Network operations	Network operations	2		
Vendor management rollout	3	Shared services/ procurement	Procurement specialist	1		MAP
		Network engineering	Network engineer	1+1		
		Network planning	Network planner	1+1		
		Product/service knowledge				
Change management rollout	4	Network planning	Network planner	5		
		Network engineering	Network engineer	3		
		Network management	Network management	1		

Note: J = job description needs to be created; T = training required; SLA = service level agreement; and H = new hire.

Table 17.6. Implementation Action Planning: Management Roles and Responsibilities

Infrastructure	*Roles and Responsibilities*
Transformation council	Corporate goals deployment Plan review, approval, and support Allocate plan resources Periodic progress review
Process owner	Lead action plan development effort Lead action plan deployment effort Review action plan progress Report results to transformation council and stakeholders
Process management team	Develop action plan Help process owner deploy action plan Monitor action plan progress Interface directly with stakeholders on plan deployment
Functional manager (stakeholder)	Help process owner and process management team deploy action plan in own functional area Suggest ideas for improvement

18 The Process-Centered Organization in Operation

In the process-centered organization, at any level, the business process is the defining organizing concept for structure, resource deployment, and management.

18.1 The Business Process Level

The individual business process essentially comprises all the work-related interactions among people who are suitably situated in a logical, physical, and geographic pattern and who operate according to an applicable set of values, policies, and practices. As in every organization, these people are supported by specified human resources subsystems.[1]

A key component of the process-centered organization is the process management infrastructure, already described in Chapter 9 (Figure 9.2). Its prime purpose is to provide the framework for managing and supporting process ownership and accountability. This statement is based on two fundamental conclusions:

1. Process adaptability can only be attained through an effective commitment management which, in turn, can only be ensured by an empowered, well functioning process ownership.
2. Process centering is essentially customer and outcome oriented. This means that it, too, is dependent on the reliable functioning of the network of commitments and associated process ownership.

Accordingly, our discussion here will focus on how the process-centered organization must operate to make this happen. From the process point of view, the key organizational mechanisms are

- Process ownership
- Network of commitments defined by the customer-supplier value chain
- Accountabilities associated with commitments
- Process stakeholdership
- Continuing process assessment

18.1.1 Process Ownership

Process ownership is established at the beginning of the reengineering project and documented in the original process contract. Through the first two phases of the project, the process owner assumes specific accountabilities for (1) delivering the completed new process plan, and (2) completing the process implementation plan with the associated project plan and budget. At the completion of each of these phases, the process owner owns the process design and its implementation.

In the third operating phase, the process owner becomes accountable for the ongoing delivery of committed process results that meet the agreed-to conditions of satisfaction. Given that the business process represents a strategic asset, the process owner can also be viewed as an asset manager. Accountabilities for partial results that contribute to the main deliverables are assumed by various individuals along the customer-supplier chain, many of whom are members of the process management team. It is crucial to the successful implementation of the reengineered process that *the continuity of accountabilities is maintained into the operating phase.*

There are three alternative approaches to transferring the responsibilities of the process owner and team into the operating phase, but only two of them preserve some level of accountability (Figure 18.1):

1. After completing the process design and implementation planning phases, the same process owner and team continue with the operational management of the reengineered process. In this assignment, the process owner and key members of the process team are engaged full time. This approach ensures the smooth transfer of process-level accountability and the full adaptability of the operating process from the outset.
2. Upon completion of the process transfer phase, only a skeleton full-time process team is maintained during the operations phase, led by a new process owner with only coordination responsibilities. Process resources are managed functionally, and accountability for partial outcomes is assumed among the functions. This approach contains a single discontinuity at the transfer point into the operations phase in that process ownership, in fact, changes hands.
3. The process management infrastructure is essentially disbanded upon having completed implementation planning. The process owner and members of the process team resume their earlier functional assignments, as all process

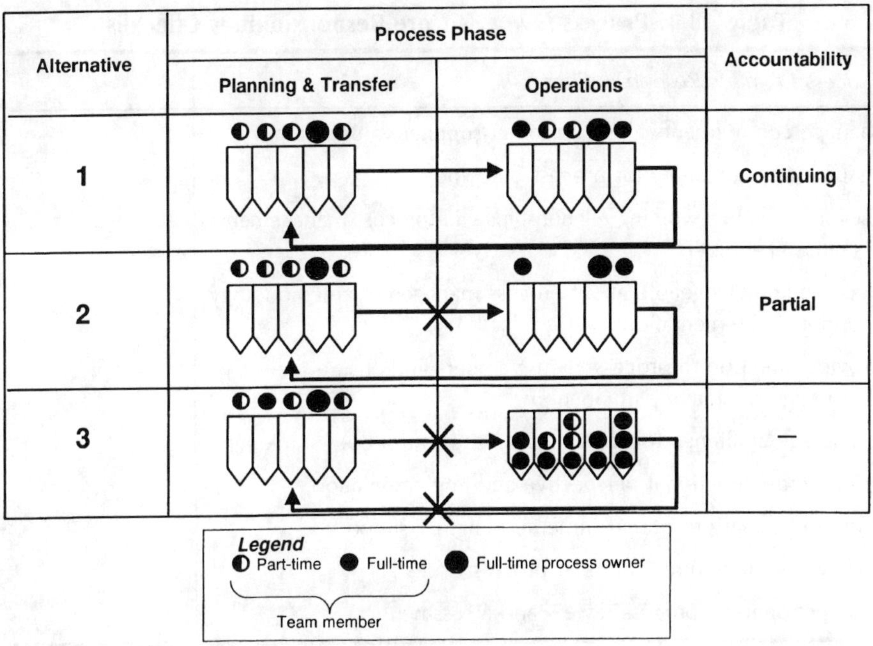

Figure 18.1. Business Process Ownership and Accountability

resources are returned to the functions. Accountability is dissipated beyond recognition. This approach contains two discontinuities: one occurs at the transfer point into operations, the other when at some point in the future a new process reengineering effort needs to be started and a new design team formed.[2] This alternative is not suited for a process-centered environment.

Smoothly functioning process ownership and accountability also provide the solution to the two dilemmas of process management, first formulated by Hammer:[3] (1) variation and dispersion, and (2) scale and complexity. Variation and dispersion are the apparent conflict between the need for process consistency across the enterprise and the variation caused by geographic dispersion and differences in local market requirements. Traditional, workflow-oriented process management tries to solve this dilemma by creating multiple, different versions of the same basic process flow, each under the leadership of a working owner, and assigning an operations director to each version to deal with variation in market (customer) requirements. The holistic approach to process management, on the other hand, maintains consistency through the sameness of the customer-supplier dialogue, while allowing variation and adaptability through the flexibility of commitment management.

Scale and complexity are the combined effect of process size (amount of resources, number of operating locations) and structure (number of constituent subprocesses and activities, hand-off points, multiple process versions). Scale and

Table 18.1. Process Owner's Core Responsibilities Checklist

Process Owner's Responsibilities	✔
Manage commitments along customer-supplier value chain	
Assume accountability for process outcome	
Establish effective working relationships with process management team/stakeholders	
Become knowledgeable about process management methodology and its implementation	
Provide direction to process team members and ensure their training and active participation	
Encourage challenge to current practices ("out-of-box" thinking)	
Ensure cross-functional perspective and implementation	
Assure necessary resources (time, funding, I/T support)	
Resolve/escalate intra- and inter-process issues	
Conduct ongoing process review and assessment	

complexity can make a single process owner's job extremely challenging. Again, the traditional approach is the subdivision of the process and the assignment of additional process management resources. Subprocess owners operate under the overall management umbrella provided by the executive process owner. The execution of each subprocess is overseen by an operations director. As we have already shown, the holistic approach is size independent. Complexity is addressed by the creation of subordinate value chains, each with its own set of customer-supplier relationships. While this may add to the size and the amount of resources in the business process, it does not create uncontrollable complexity.

To ensure the smooth functioning of ownership, the process owner must discharge a number of core responsibilities, shown in Table 18.1. First of all, to succeed in this, the owner must have clout (that is, authority and control over resources) and empowerment to make decisions and negotiate commitments. Second, the personal attributes of the process owner must include process thinking, foresight, openness to new ideas, strong communication skills, persuasiveness, determination, and result orientation.

18.1.2 Accountability Framework

The accountability framework of the business process is defined by the set of suppliers committed to the satisfactory delivery of a sequence of related process outcomes within a given business cycle. Because outcomes are driven by commitments, the accountability framework, by definition, includes those suppliers linked to the backbone network of commitments; however, it also covers that part of the customer-supplier dialogue which contains the commitments needed to produce the overall process outcome.

Using the human body as a rough analogy, the backbone network can be viewed as the skeleton. The accountability framework is the entire system of motor nerves which, supported by the skeleton, can activate the muscles to perform a specific task. Accordingly, the accountability framework is characterized by:

- *Boundaries:* The length of the customer-supplier value chain, which then also defines the number and extent of customer-supplier dialogues within it.
- *Process professionals:* Perform work as individuals or as members of teams. They are all customer driven and capable of solving problems and taking responsibility on their own initiative. Their personal characteristics include process thinking and result orientation; they possess an education and knowledge base that prepares them to deal with the unexpected and assume accountability, as required.
- *Teams:* Work along the value chain and across functions and hand-off points. Functional teams, such as self-directed work teams or quality circles, are part of the procedural workflow.
- *Reporting lines:* A series of peer relationships reflected by the customer-supplier relationships and the process team structure; however, traditional reporting lines may be retained along the hierarchically arranged layers of work units (activities, tasks, work steps) within functional segments of the workflow.

18.1.3 Process Stakeholders

From an organizational viewpoint, the significance of process stakeholders lies in their contribution to the integrity of commitment management and to the availability of adequate process resources. The rationale behind process stakeholders, of course, is to secure the support of every functional head in the organization who has resources in the business process to be reengineered or who will be affected by it. In addition, these functional executives may have valuable inputs for process-related decision-making; in many instances, their cooperation or influence on others may be vital to the successful completion of the reengineering project. For these reasons, the process owner will make sure stakeholders of the process are kept informed throughout the reengineering effort; their concerns, if any, will have been addressed and their commitment is always ensured.

The reengineering of Pfizer Central Research Division's procurement process could not have become the success it was without the thoughtfully planned and executed stakeholder participation in the management of the reengineering effort. To this day, process stakeholders continue to be involved in the ongoing operational management of the reengineered way of procuring the myriad of supplies critical for the company's research function.

Risks	Process Management Action
Instability; out-of-control state	Corrective action
Process performance decline; ineffectiveness; inefficiency	Problem solving; improvement projects
Non-competitive process	Periodic assessment

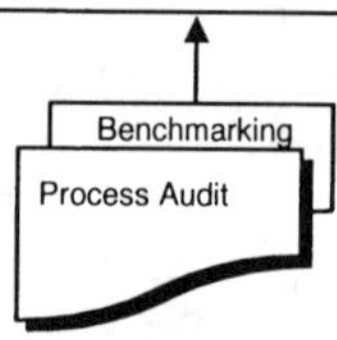

Figure 18.2. Continuous Process Improvement and Assessment

A similar example of effective stakeholder relationships and ongoing active participation is the order-fulfillment process at Union Camp's Fine Paper Division. From the very beginning of the reengineering effort, process stakeholders formally committed themselves to specific roles and responsibilities throughout the project. Their involvement was secured through scheduled monthly stakeholder meetings, as well as personal participation in key events during the process design phase and, in particular, during implementation planning.

18.1.4 Continuous Process Assessment

Continuous process assessment provides the process owner with a reliable base of reference for process effectiveness (product quality) and efficiency (lowest possible resource consumption) on an ongoing basis (Figure 18.2). From this base, the owner can negotiate meaningful and realistic commitments in the face of change. This ability of the owner is the prerequisite for process adaptability.

Even when there is no external change, periodic process self-assessments will still provide the impetus for process redesign or even reengineering. These assessments may take a variety of forms. The scheme shown in Figure 18.3 was originally developed by IBM for the simultaneous assessment and formal corporate audit of its internal business processes (Figure 18.4). In this approach, performance measurements and other process indicators are captured in a flat database. During the audit, the audit system simply "plugs" into the database and determines (1) if the process is audit-ready, and (2) its assessed status. This scheme has endured because it does include adaptability as the most advanced process attribute to aim for.

Rating	Assessment Criteria			Basics of Process Management						Continuous Improvement				
		Process Identified	Process Owner Named	Management Infrastructure In Place	Process Defined & Documented	Input/Output Relationships	Measurements/Control	Customer Requirements	Quality Control Methodology	Control	Improvement	Effectiveness	Efficiency	Adaptability
5	Process identified/ownership assigned	X	X											
4	Process manageable and stable	X	X	X	X	X	X	X	X					
3	Process effective	X	X	X	X	X	X	X	X	X	X	X		
2	Process efficient and competitive	X	X	X	X	X	X	X	X	X	X	X	X	
1	Process adaptable; "best in class"	X	X	X	X	X	X	X	X	X	X	X	X	X

Figure 18.3. Process Assessment Criteria

Another process assessment model, originally developed for the software industry by the Software Engineering Institute (SEI) is the Capability Maturity Model (CMM).[4] It is a conceptual framework that represents the software development process and its management. Its purpose is not so much to evaluate current performance, but to assess the capability of the software development process. Not unlike the IBM process assessment scheme, CMM defines process capability in terms of five maturity levels (Figure 18.5):

1. *Maturity Level 1: Initial State.* The software process can be characterized as *ad hoc*, undefined, or even chaotic. Even as of this writing, the software development processes of many industrial and business organizations still fall into this category.

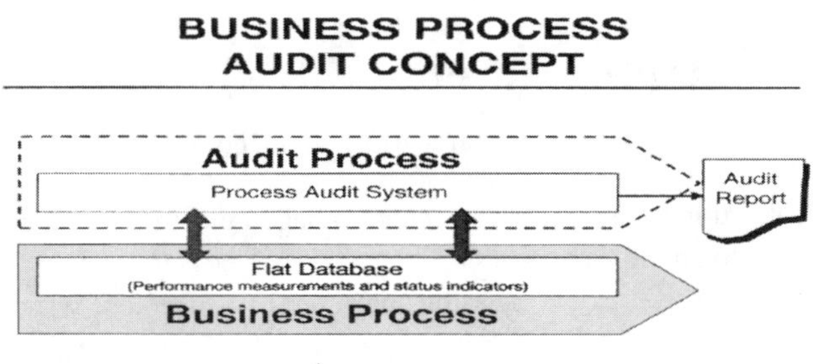

Figure 18.4. Business Process Audit Concept

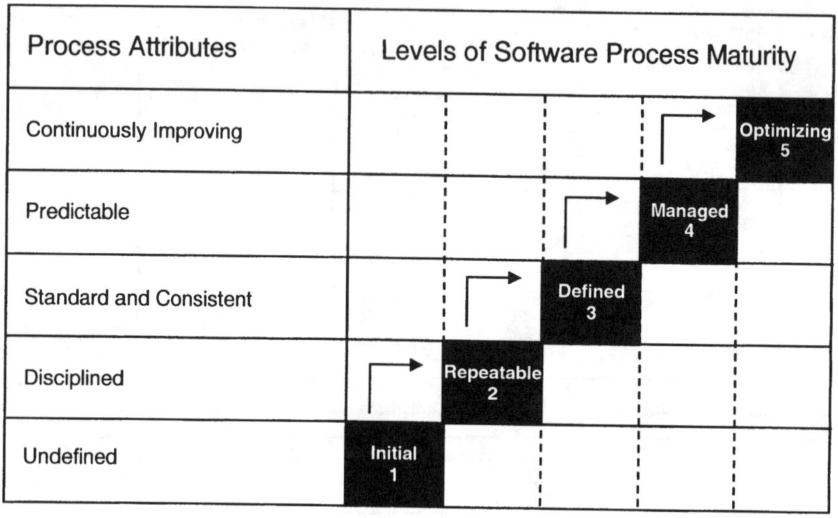

Figure 18.5. The Capability Maturity Model

2. *Maturity Level 2: Repeatable State.* The process is disciplined.
3. *Maturity Level 3: Defined State.* The process is consistent.
4. *Maturity Level 4: Managed State.* The process performance is predictable.
5. *Maturity Level 5: Optimizing State.* The process is continuously improving and is the equivalent of a fully implemented total quality management (TQM) system.

These five levels of maturity reflect that CMM is a model for both assessing and improving the capability of software processes along a continuous scale, from the initial primitive state all the way to measurable, optimized performance. While originally developed for the software development community, currently ongoing efforts to produce new, generic versions of CMM should provide process managers with a disciplined and powerful tool.

18.2 The Enterprise Level

Elements of the process-centered organization at the enterprise level include:

- The collection of individual process management infrastructures
- The transformation council
- A comprehensive, balanced, value-oriented measurement system that maximizes overall enterprise performance
- The extended accountability framework that encompasses the senior management of the enterprise

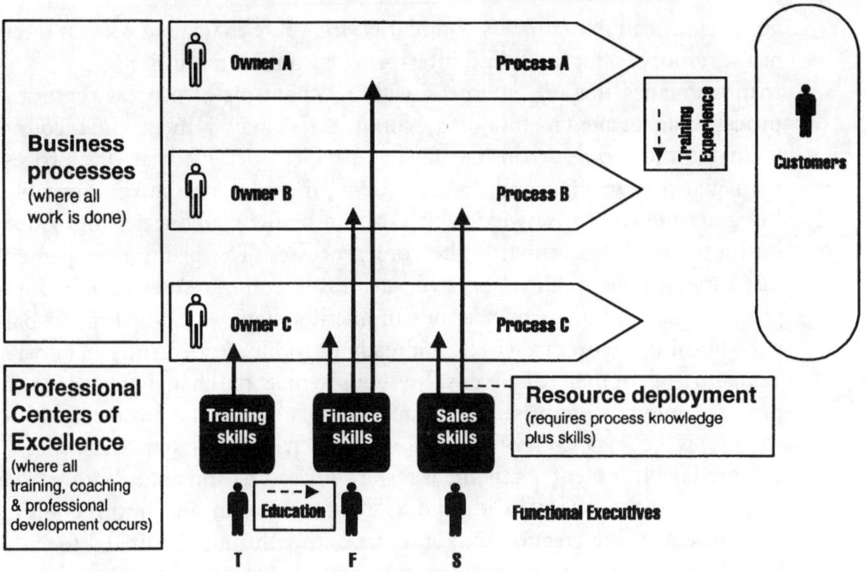

Figure 18.6. The Process-Centered Organization

18.2.1 Operational Responsibilities

Combining several individual process infrastructures allows the development of an early form of the process-centered organization (Figure 18.6). It is a structure where all work is done in the business processes led by their respective owners. The management of people, on the other hand, is carried out by functionally oriented executives in *professional centers of excellence.* It is these centers where all functional training, professional development, and individual coaching are done. Identification with a specific center is determined by the individual's basic education; assignment to a specific business process is decided on the basis of professional knowledge, experience, and specific skills acquired through training.

The purpose of the enterprise, above all, is value creation — more specifically, providing customer value. Customer value is created in the enterprise's business processes; competitive advantage derives from the superiority of these processes. Enterprise management, through the transformation council, exercises a number of key responsibilities related to the reengineering of individual business processes and the building of the process-centered enterprise:

1. Take actions against the fragmentation around business processes, and prevent the formation of "horizontal stovepipes" — which are the opposite of what is intended by process centering. These actions include the coordination of process goals and their alignment with the organization's strategic

goals; clarification of process boundaries, the value chain, and process interfaces; resource allocation; and inter-process conflict resolution.

2. Institutionalize the key enterprise-wide mechanisms of process centering: process management methodology, commitment management, and accountability framework. For the satisfactory performance of these, the process management executive and the individual process owners are accountable.
3. Integrate the enterprise-wide value creation by the reengineered individual business processes, primarily the core processes. The prime management instruments here are the integrated business planning system, and the integrated enterprise performance measurement system, embodied in the balanced business scorecard (BBSC), already introduced in Chapters 9 and 12.
4. Transform all human resources subsystems to meet the changing needs of the process-centered enterprise. The prime candidate (and the most difficult to change) is the reward-and-recognition (compensation) system. In a process-centered environment, its theme is reward for results and contribution. The alignment of the enterprise around its key processes means, in this context, that process value creation and enterprise growth must be linked to individual employee performance measurements, incentives, and recognition.

Process centering the enterprise also means the upward extension of the accountability framework to cover the senior management of the organization. Just as the accountability for business process-level outcomes and performance rests with the respective process owners, so the ultimate accountability for enterprise-level outcomes (profit, shareholder value, market acceptance, growth, and so on) must be assumed by one individual: the chief executive officer. These cascading accountabilities start with the CEO — who is, in most cases, the transformation executive — and extend all the way into the flow of individual processes (Figure 18.7).

18.2.2 The Millennium Enterprise

The arrival of the 21st century has been used by politicians, pundits, and publicists as a metaphor for challenge, change, and renewal. Among its various characterizations have been included the quality century, the process century, the *e*-century, and, more recently, the virtual century. Whichever designation we feel most comfortable with, one thing is clear: The 21st century will be very, very different from the last one.

The 1980s and 1990s have already brought us an information explosion that ushered in the information age and with it such phenomena as world-wide networking, *e*-business, and the information economy.

The management of most organizations has traditionally sought to achieve better business performance and competitive superiority through a deliberate customer orientation:

- Customer satisfaction achieved by means of quality products and services
- Improved productivity through cost and cycle time reduction.

Results		Transformation Executive	Process Owner	Process Team Member	Individual Professional
Enterprise	Outcome	(✓)			
	Performance	✓			
Business Process	Outcome		(✓)		
	Performance		✓		
Subprocess or Activity	Outcome			(✓)	
	Performance			✓	
Task or Work step	Outcome				(✓)
	Performance				✓

(✓) = prime commitment driven

Figure 18.7. Accountability Framework

The key task of 20th century executives and managers has been performance improvement. The tools to achieve this were supplied by total quality management (TQM), business process reengineering (BPR), and organizational effectiveness (OE). The solutions of TQM and BPR were process solutions that focused primarily on key, individual business processes. Now, management faces the compelling requirement of doing all this continuously, faster, and better than the competition — and in the face of accelerating and unpredictable change. The double challenge of the 21st century is to be successful while continuously adapting and responding to a highly competitive, fast-moving, information-driven, and nearly chaotic environment. In fact, "continuous adaptation" is the motto for the new century.

The truly adaptive enterprise of the new century will be process centered, aligned around its adaptable core processes (Figure 18.8). It will be managed as an open system, with the aid of a balanced planning and measurement system, along the lines of the balanced business scorecard. In this enterprise, benchmarking takes on a new meaning. Instead of comprising a series of finite projects, it becomes the information gathering and knowledge capturing sensor of the organization. The massive flows of information will be managed by enterprise resource planning (ERP) systems — extended to the suppliers and customers of the enterprise, supporting a true implementation of *e*-business. The organization's customer orientation will be ensured by a pervasive TQM infrastructure. The prime operational management tool will be process management, and the prime form of doing business will be the integrated value chain.

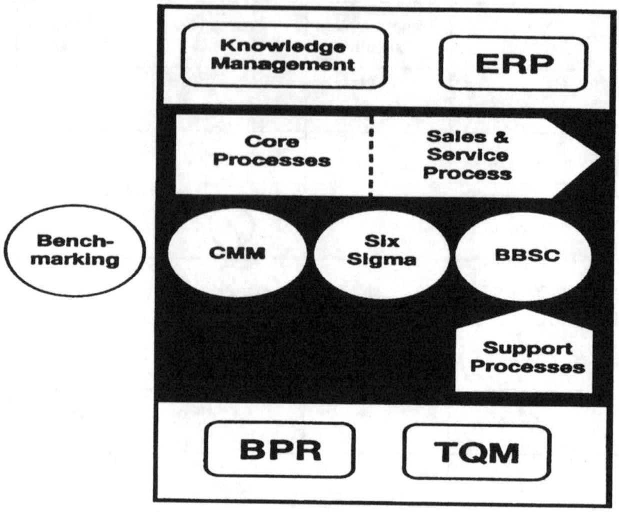

Figure 18.8. The Millennium Corporate Business Model

In such an enterprise, there cannot be differences in the performance of various business processes. Just as a successful company cannot afford unreliable suppliers, the core processes within the enterprise will not achieve their ambitious performance goals without comparable performance by support and infrastructure processes. The tools for tuning process performance are Six Sigma for performance monitoring and improvement and the Capability Maturity Model for disciplined and consistent process assessment.

Large or small, this millennium enterprise will be entrepreneurial, highly integrated and finely tuned in its goals, decision-making, commitment management, and performance measurements. Competitive success does not come from lofty goals and ambitious objectives alone. It will come from the disciplined and highly integrated implementation of a number of interrelated management concepts and methods — which all add up to the process-centered enterprise of the millennium.

18.3 Summary

In the process-centered organization, at any level, the business process is the defining organizing concept for structure, resource deployment, and management. Once it becomes operational, the newly reengineered business process must be adaptable from the very beginning. One prerequisite for this is the seamless transfer of the process owner's accountability into the operations phase. The other is an empowered, well-functioning process ownership. The effective functioning of process ownership, in turn, depends on several organizational mechanisms, such as the accountability framework, continuous process assessment, and process stakeholdership.

The accountability framework of the business process is defined by the set of suppliers committed to the satisfactory delivery of a sequence of related process outcomes within a given business cycle. Smoothly functioning process ownership and accountability also provide the solution to the two dilemmas of process management: (1) variation and dispersion, and (2) scale and complexity.

Continuous process assessment provides the process owner with a reliable base of reference for process effectiveness (product quality) and efficiency (lowest possible resource consumption) on an ongoing basis. From this base, the owner can negotiate meaningful and realistic commitments in the face of change. This ability of the owner is the prerequisite for process adaptability. Even when there is no external change, periodic process self-assessments will still provide the impetus for process redesign or even reengineering.

The truly adaptive, entrepreneurial enterprise of the millennium will be process centered, aligned around its adaptable core processes. It will be managed as an open system, with the aid of a balanced planning and measurement system. In this enterprise, benchmarking it becomes the information gathering and knowledge capturing sensor of the organization. The massive flows of information will be managed by enterprise resource planning systems — extended to the suppliers and customers of the enterprise, creating an integrated value chain for the true implementation of *e*-business. The organization's customer orientation will be ensured by a pervasive TQM infrastructure. The prime operational management tool will be process management, supported by advanced methodologies such as Six Sigma for performance monitoring and improvement and CMM for consistent process assessment.

Appendixes

Appendix 1. The Tools of Process Centering

There is a virtually infinite number of tools and techniques currently available or under development that, in some way, perform functions useful or necessary for process centering; therefore, it would be a daunting task to try to present an even partial listing of what is available. (The annual directory of software implementations of quality and process management-related tools published in *Quality Progress* now runs to 23 pages listing only the suppliers of such tools by application category, but not the tools themselves.) Instead, our purpose here is to present a structured framework of the available tools with a sampling of those that, in our experience, are the most frequently used or better known examples of their category. The appearance of any particular tool in this listing does not represent an endorsement of one tool over another possible alternative.

It is convenient to divide the tools of process centering into two major categories:

1. *Stand-alone tools* are designed and packaged to perform a single or specific set of related applications. These tools may operate under one or more operating systems and can be acquired and installed as a stand-alone package, hence the name.
2. *Enterprise resource planning (ERP)-dependent tools* may perform similar application functions as the stand-alone tools but are embedded in an ERP system (SAP, Baan, Oracle, PeopleSoft, etc.) and can be obtained only as part of that enterprise system solution.

As more and more organizations are implementing some form of ERP systems, it is important that executives and managers are aware of this distinction and make their choices with a full understanding of the alternatives as well as potential overlaps.

A1.1 Stand-Alone Software Tools

A1.1.1 Process Modeling Tools

Process modeling tools include simple flowcharting/flow diagramming techniques as well as more sophisticated process map generators and description methods. Process modeling tools provide the capability to represent an organization by mapping or flowcharting its cross-functional business processes. Modeling tools allow the process owner or team to define a business process in terms of a series of process maps (flow diagrams) of varying detail. Process maps show, in graphic form, both the flow of work and information in a process and allow process professionals to walk through a number of process alternatives in an effort to determine how a process will run at optimal performance. Modeling software tools also allow the capture and tracking of changes to a process design and help determine where measurement and control points should be placed and where opportunities arising from waste elimination, cycle time reduction, and other improvements exist. Table A.1 shows a listing of the most widely used and, to our way of thinking, the more advanced process modeling tools currently available.

Process mapping tools known as IDEF are almost a separate category among process modeling tools. Originally developed under the sponsorship of the U.S. Air Force, integrated computer-aided definition (IDEF) is a process-mapping technology for performing basic process management functions such as process analysis, cost/benefit analysis, requirements definition, functional analysis, systems design and maintenance, and baseline process documentation — in support of a business process reengineering effort. IDEF process maps comprise three types of information: graphic, text, and glossary. The most frequently used IDEF process mapping products are included in Table A.1.

A1.1.2 Process Documentation Tools

Process documentation tools are used in all three phases of process management: design, implementation planning, and operations. They are most valuable during (1) process definition, when little or nothing is known about a given business process, and (2) in operations, as user guides and documentation. The tools listed below facilitate the creation of process documentation. They provide a structured and high-quality framework for creating the initial document as well as the control, document migration, and document integration.

The best examples of process documentation tools include text processors such as Microsoft's *Office 97*, Novell's *WordPerfect*, both running under Microsoft Windows 95 (and later versions) and *ClarisWorks* (on Macintosh computers). Other prominent examples of software tools specifically developed for process documentation are *Docsite* by Insite Objects, Inc., and *Imap* by Documentation Associates (see also Table A.1).

A1.1.3 Process Simulation Tools

Simulation is an exercise that substitutes or artificially devises some components or parameters found in the real-life process to allow process owners and teams to analyze, practice, or observe process activities, tasks, or work steps. Generally speaking, process simulation tools are used to examine the dynamic behavior of processes in a "what if" and "to be" context. Simulation is usually carried out in conjunction with an already-existing process map representation of the existing or desired process. Simulation provides an effective way of examining potential process changes before resources are invested in the actual effort. Simulation also provides an early warning of implementation errors, conflicts, or disconnects without actually making the planned change in the real-life process.

Simulation techniques are usually divided into two classifications: discrete and continuous. Because most business process maps are a collection of discrete tasks or work steps, the more popular and widely used process simulation tools employ discrete simulation techniques. The most prominent simulation tools are listed in Table A.1.

An emerging variation of simulation called virtual reality (VR) uses automated devices to create an artificial representation of a real environment. The VR devices contain sensors, actuators, and display devices which are usually attached to or worn by participants. The devices sense the participants' movements and respond in real time by providing the appropriate sensory feedback — changing graphic displays or resistance to physical movement. While VR technology is still somewhat unwieldy and expensive, it is increasingly popular in the testing, management, and operation of some very complex processes, such as flying a combat airplane or performing complex surgical or other medical procedures.

A1.1.4 Process Mapping-Related Activity-Based Costing Tools

Activity-based costing (ABC) is one of the most widely used methods to determine resource consumption by a given business process in terms of time and cost. ABC defines and manages product and process costs more effectively than the traditional ledger-based cost accounting approach. It is a proven and widely used methodology that determines the cost performance of work activities and related resources and cost objects. ABC also recognizes the causal relationship between cost drivers and the activities performed in a given process.

Business process reengineering and, specifically, analysis in terms of economic value added (EVA) have refocused attention on the contribution of process activities to improving the value received by the customers of the process, as well as the value created by the process for the organization itself. Examples of the most popular and widely used ABC software tools are provided in Table A.1.

Table A1.1. Software Tools of Process Centering

Methodology/ Category	Software Tool	Platform/ System	Developer/Vendor Name	Address	Telephone/ Fax No.	http//:www.
Process modeling	Flowcharter 7	IBM PC 486+ MS Windows 95+	Micrografx	1302 E. Arapaho Richardson, TX 75081	972-234-1769 972-994-6694	micrografx.com
	Flow Model	IBM PC 386/higher MS Windows 3.1+	Qsoft Solutions	445 W. Commercial St. E. Rochester, NY 14445	716-264-9700 716-264-9702	qsoftguide.com
	Livemodel	SAP R/3	Intelli Corp	1975 El Camino Real W. Mountain View, CA 94040	650-965-5500 650-965-5647	intellicorp.com
	Process Charter	IBM PC 386+ MS Windows 3.1+	Scitor Corp.	333 Middlefield Rd. Menlo Park, CA 94025	415-462-4200 415-462-4201	scitor.com
	VISIO	IBM PC 486+ MS Windows 95+	Visio Corporation	520 Pike Street Seattle, WA 98101	206-521-4500 206-521-4501	visio.com
	ARIS 4.0	SAP R/3	IDS Scheer	2 Brandywine Chaddsford, PA 19317	610-558-7600 610-558-7603	ids-scheer.com
	Design/IDEF	IBM PC 486+ MS Windows 95+ Macintosh Macintosh OS	Meta Software	125 Cambridge Park Dr. Cambridge, MA 02140	617-576-6920 617-661-2008	metasoft.com

Process documentation	ISOlution 3.0		Insite Technology Group	7055 Engle Road Cleveland, OH 44130	800-446-8665 216-826-3431	insite-inc.com
	MS Word	IBM PC 386+ MS Windows 95+	Microsoft Corp.	One Microsoft Way Redmond, WA 98052	425-882-8080 425-936-7329	microsoft.com
Process simulation	PROMODEL	IBM PC 486+ MS Windows 95+	PROMODEL Corp.	1875 South State Orem, UT 84058	801-226-6036 801-226-6046	promodel.com
	Work Flow Analyzer Design/CPN	IBM PC 486+ MS Windows 95+	Meta Software	125 Cambridge Park Dr. Cambridge, MA 02140	617-576-6920 617-661-2008	metasoft.com
	WITNESS	IBM PC 486+ MS Windows 95+	AT&T ISTEL	25800 Science Park Dr. Beachwood, OH 44122	216-292-2668 216-292-2861	
Process-related ABC	Easy ABC Plus	IBM PC 486+ MS Windows 95+ OS/2	ABC Technologies	5075 S.W. Griffith Dr. Beaverton, OR 97005	503-626-4895 503-626-4003	abctech.com
	Easy ABC Quick	IBM PC 486+ MS Windows 95+	Qsoft Solutions	445 W. Commercial St. E. Rochester, NY 14445	716-264-9700 716-264-9702	qsoftguide.com
Project management	MS Project	IBM PC 486+ MS Windows 95+	Microsoft Corp.	One Microsoft Way Redmond, WA 98052	425-882-8080 425-936-7329	microsoft.com

A1.1.5 Project Management Tools

Whether it is a simple incremental process improvement or the reengineering of an entire, complex, cross-functional business process, the implementation of such a change usually results in a process implementation project. Such projects state objectives in terms of schedule, cost, and quality, just to name the three key project parameters. Project management techniques range from the use of simple Gantt charts representing project schedules all the way to complex resource, time, and quality management systems encompassing critical path methods (CPM), project evaluation and review techniques (PERT), and the like. Accordingly, project management software tools include basic project management packages provided by major software vendors such as Microsoft's *MS Project* as well as sophisticated stand-alone packages, several of which are listed in Table A.1.

A1.1.6 Groupware/Software Tools for Team Effectiveness

Team-oriented software tools are designed to facilitate (1) training or knowledge transfer to process teams, and (2) implementation of process work itself by teams. Suppliers of teamwork-oriented software tools include CEBOS (www.cebos.com), Context Integration (www.context.com), Information Technologies (www.infotech-usa.com), MicroMedia (www.qskills.com), OXKO Corporation (www.oxko.com), and Training Technologies (www.surveytracker.com).

A1.2 ERP-Based Software Tools

Enterprise resource planning systems include mainline software which comprises the functionality of major business processes such as materials management, production planning, accounting, controlling, sales and marketing, and customer service. Along with this mainline software, ERP vendors have created a set of supporting software tools (integrated in the ERP package) for the operational implementation of the various functional software modules and their deployment in customized environments. These tools, which are usually part of the ERP installation, are essential during the implementation effort as well as in the actual operation of the ERP system.

In general, ERP-related software implementation tools have three objectives:

- Description of the functionality and business processes encompassed by the ERP solution
- Configuration of the enterprise software modules
- Implementation planning and scheduling

A1.2.1 SAP

The SAP R/3 enterprise resource software is supported by the *Business Engineering Workbench* (BEW), which is the basic set of tools used for configuring the R/3 software modules. (Release 4.0 and higher releases of R/3 will be supported by a more advanced tool package known as the *Business Engineer.*) The main components of the BEW are

- *Procedure Model:* A general overview of SAP's procedures for implementing the enterprise solution. It covers the life cycle of a system implementation in its four phases (organization and conceptual design, detailed design and implementation, preparation for production, and operations). The procedure model is also complementary with other SAP and third-party implementation methodologies. Finally, the procedure model also covers systems maintenance and enhancement procedures, such as release upgrades.
- *Reference Model/Process Model:* Demonstrates the functionality and business processes included with SAP. The process model provides five views of a given function or process:
  1. Process
  2. Data and information
  3. Distribution model
  4. Business object model (such as products, customers, facilities, cost centers, etc.)
  5. Organization

  Of these, the most commonly used view is that of the process, which allows the representation of a series of business events in terms of process steps and activities. The process view can also be the basis for the application of stand-alone process modeling tools that are SAP compatible. Examples of process modeling tools compatible with SAP include *Life Model* by Intellicorp, *ARIS* by IDS Scheer, and *VISIO* by Visio Corporation.
- *Implementation Guide (IMG):* An online configuration aid providing tables and activities required to configure the enterprise software system. IMG is also a tool for documenting, validating, and controlling the software configuration.

A1.2.2 Oracle

The process modeling software tool supporting the Oracle enterprise resource software is a diagrammer, known as the *Business Process Modeller* (BPM). It resides in the Designer/2000 component of the Oracle software package. BPM displays processes, data flows, and data stores, as well as the organization units using them. It lets the user diagram the data flowing between processes and data stores. BPM also makes it possible to "drill down" from a higher level process to a lower level and diagram the subprocesses that compose the overall process. Last, but not least, BPM can be used to show external triggering events and outcomes to processes within an organization. It can be used to conceptually show and possibly redefine what happens in the current process, or what would happen in the new process. As such, it is an effective process reengineering tool.

Appendix 2. Abbreviations and Acronyms

Abbreviation/Acronym	*Explanation/Description*
ABC	Activity-based costing
ABM	Activity-based management
BPQM	Business Process Quality Management® (Juran Institute's methodology and workshop for BPR)
BPR	Business process reengineering
CAD	Computer-aided design
CAM	Computer-aided manufacturing
CAPP	Computer-aided process planning
CASE	Computer-aided software engineering
CBAM	Cost/benefit analysis methodology
CEO	Chief executive officer
CFO	Chief financial officer
CIO	Chief information officer
CMM	Capability Maturity Model
COQ	Cost of quality
COQA	Cost of quality analysis
CPM	Critical path method
CQO	Chief quality officer
DOE	Design of experiments
EDI	Electronic data interchange
ERP	Enterprise resource planning (systems)
EVA	Economic value added
FMA	Failure mode analysis
FMECA	Failure mode, effects, and criticality analysis

IDEF	Integrated definition language
IPO	Input-process-output (process model)
IS or I/S	Information systems (or services)
ISO	International Standards Organization
IT or I/T	Information technology
JIT	Just-in-time
LCC	Life-cycle cost
MRP	Material requirements planning
MRP II	Material resource planning
PDCA	Plan-do-check-act (cycle)
PERT	Project evaluation and review technique
QA	Quality assurance
QC	Quality control
QFD	Quality function deployment
QI	Quality improvement
QP	Quality planning
ROA	Return on assets
ROCA	Return on controllable assets
ROE	Return on equity
ROI	Return on investment
RONA	Return on net assets
SPC	Statistical process control
SQC	Statistical quality control
SQP	Strategic quality planning
TCM	Total cost management
TQM	Total quality management

Glossary

Glossary

Accountability: personal acceptance of the consequences, be they reward or retribution, of making a commitment.

Accuracy: closeness of the average of a series of repeat measurements to the accepted true value. The smaller the difference between measurements and the true value, the more accurate the measuring device.

Activity: collection of tasks performed to change inputs into outputs. A small process that is part of, but subordinate to, a subprocess. A special type of activity is a communication act, where specific information (input) is conveyed in verbal or symbolic (written) form (output) to a recipient. *See also* subprocess.

Activity-based costing (ABC): a method of gathering data and establishing the actual cost of a business process by determining the resource consumption of all the activities in the process. When the method targets tasks, it is known as task-based costing.

Adaptability: ability to achieve higher and improved levels of capability in the face of change. An adaptable process maintains its effectiveness and efficiency in the face of change over time.

Adaptive loop: procedural implementation of the four functional steps of (1) observation or sensing, (2) orientation or identification, (3) decision, and (4) action. Its different versions are also known as the feedback loop or control loop.

Appraisal: act of judging as to quality, status, or value.

Availability: probability that a product or service, when used under specified conditions, will perform satisfactorily every time it is called upon.

Block diagram: a form of a flow chart, showing a series of work activities and the relationships among them (as supplier-customer), but without showing inputs, outputs, or information flow.

Business process: in general, a major cross-functional process producing an output (goods or services) of importance, for whose management a permanent infrastructure has been (or is being) set up. *See also* holistic process model, process.

Business process owner: *see* process owner.

Capability: inherent uniformity of outcomes that can be produced by resource-consuming process components performing specific work activities.

Capability Maturity Model (CMM): (1) a structured, conceptual framework that defines the management of software development and maintenance processes; (2) a management tool for software and I/T organizations to systematically improve their development and maintenance processes.

Cause: an established reason for the existence of an outcome.

Cause-and-effect diagram: a problem-solving tool based on the graphic representation of the causal relationships among process components used to analyze the potential causes of variation. Also referred to as fishbone diagrams or ishikawa diagrams.

Commitment: negotiated agreement between two parties to produce a defined outcome and accept it if produced within stated conditions of satisfaction.

Common cause: a source of variation in the process output that is inherent to the process and will affect all the individual results or values of process output.

Communication act: one of seven possible, unidirectional transfers of information between two parties, usually a customer and supplier. Communication acts include request, offer, commit, report, accept, reject, and withdraw. Also known as speech act.

Communication protocol: description of the seven communication acts, along with the terms and conditions of their use. The communication component of the customer-supplier dialogue.

Conditions of satisfaction: an agreed-upon set of terms and conditions subject to which the customer will accept the process output. The basis for the supplier's commitment.

Continuous improvement: repeated implementation of successive cases of control and improvement.

Control: set of activities employed to detect and correct variation in order to maintain or restore a desired state; a past-oriented approach to quality management.

Control loop: a special version of the adaptive feedback loop to monitor a single control variable and provide the ability for corrective action.

Control point: specific points within the process flow where corrective action can be applied that will affect process performance. Each control point contains one or more complete control loops. Control points can be used to prevent process instability or product defects. *See also* measurement point.

Correction: totality of actions to minimize or remove variations and their causes.

Cost of failure: costs incurred in correcting defects or using defective work products; cost associated with things that have been found not to conform to requirements, as well as the evaluation, disposition, and public relations aspects of such failures.

Cost of nonconformance: costs which would disappear if all products and processes were free of deficiencies; also known as the cost of poor quality.

Cost of poor quality: *see* cost of nonconformance.

Cost of prevention: costs incurred in preventing defects and to "do the job right the first time." Cost associated with actions to minimize the cost of nonconformance (also known as the cost of poor quality).

Cost of quality: the sum of costs of prevention, appraisal, and failure. The key financial measurement tool which ties process control and process optimization into a total process management effort. It can be used both as an indicator and a signal for variation (more often, patterns of variation), as well as a measure of productivity and efficiency.

Customer: the most general and consistent definition of anyone who is the recipient or the beneficiary of the output (work product) of a process or is affected by the product or process; the recipient who must be satisfied with the output or a related outcome. *See also* external customer, internal customer, and recipient.

Customer-supplier dialogue: totality of communications and procedural activities required to produce a committed outcome (execute a single business cycle or transaction) over time. The customer-supplier dialogue comprises four phases: definition, negotiation, performance, and assessment.

Customer-supplier relationship: formal description of a single customer-supplier pair within a business process.

Customer-supplier value chain: totality of customer-supplier pairs required to produce a committed outcome.

Cycle time: elapsed time between two specified events in time.

Data: information or a set of facts presented in symbolic (descriptive or graphic) form.

Decomposition: partitioning or breaking down of a process entity (process, subprocess, activity, etc.) into its component subentities.

Defect: any state of nonconformance to requirements; synonymous with deficiency.

Detection: quality control concept based on inspection and test to detect and remove defects prior to releasing the product or service to the customer(s).

***e*-business:** The conduct of business (all transactions and processes) along a value chain including the suppliers and customers of the enterprise, integrated by means of network technology.

***e*-commerce:** The marketing, buying, and selling of goods and services over a network.

***e*-trade:** The marketing, offering, buying, and selling of securities, currency, financial products, and other instruments of investment over a network.

Economic value added (EVA): after-tax cash return on invested capital, less the cost of that capital.

Effectiveness: a process characteristic indicating that the process output (work product) conforms to requirements.

Efficiency: a process characteristic indicating that the process produces the required output at a minimum level of resource consumption; as a quantitative measure, the reciprocal of productivity.

Empowerment: providing the ability (skills and competencies), authority, and means (information, methods, and capabilities) required to reach decisions and agreements, to make commitments, and to take certain actions.

End user: *see* ultimate user.

External customer: customer who is not a member of the organization or process which produces the output. *See also* customer.

Failure cost: *see* cost of failure.

Feedback: communication of changes or performance to sources which can take appropriate action.

Feedback loop: *see* adaptive loop, control loop, and learning loop.

Fishbone-diagram: *see* cause-and-effect diagram.

Flow diagram: graphic means for depicting the flow of work steps, tasks, or activities and related descriptive information in a process.

Goal: an outcome or quantified result which one intends to accomplish or attain, with sustained effort and energy directed to it over a longer period of time.

Holistic process model: the process definition that includes the customer-supplier communication protocol and the procedural flow; the communications and activities that make up the customer-supplier dialogue.

IDEF: the abbreviation of integrated computer-aided definition language, a process modeling technique comprising graphic and narrative symbols and rules.

Improvement: organized creation of beneficial change. *See also* continuous improvement.

Input: assets required for the process to complete the activities necessary to produce a specified output (work product); that which the process transforms into output.

Interface: shared boundary between two or more processes across which objects (material) and information (communications, data) are passed.

Internal customer: customer who is affected by the product or process and who is a member of the organization or process which produces the output. *See also* customer.

Mapping: the activity of creating a flow chart or process map of a business process, showing its inputs and outputs, the sequence of work activities, and the flow of information within the process.

Measurement: objective method(s) employed to determine, in a quantified form, the actual status of a process variable or product attribute.

Mission: the purpose for the existence of an organization, enterprise, or process.

Need: lack of something requisite, desired, or useful; a condition requiring provision or relief. Usually expressed by users or customers.

Objective: an outcome or result which one intends to accomplish or attain, toward which a quantified effort is to be expended over a predetermined period of time.

Offering: specific combination of products and services provided in a "packaged" form to satisfy certain specialized needs. (The products in this context are sometimes also known as deliverables.)

Output: specified end result (work product) or outcome of the process, usually required by the recipient (customer); that which the process transformed input into.

Pareto diagram: problem-solving tool based on the ranking of all problems or sources of variation according to their relative contribution to some specified effect (outcome), such as cost or total variation. *See also* Pareto principle.

Pareto principle: in any population which contributes to a common effect (outcome), a relative few of the contributors account for the bulk of that effect. An important prioritization method, named after Vilfredo Pareto (1848–1923), an Italian economist and sociologist. *See also* Pareto diagram.

Plan: a specific course of action designed to attain a stated objective.

Policy: a statement of principles and beliefs, or a settled course, adopted to guide the overall management of affairs in support of a stated mission. It is mostly related to fundamental conduct and usually defines a general framework within which other business and management actions are carried out.

Precision: the closeness of a group of repeat measurements to a mean value. The smaller the difference between the group of repeat measurements and the mean value, the more precise the instrument. Precision is an indicator of the repeatability, or consistency, of the measurement.

Problem: a question or situation proposed for solution; a state of want or need requiring remedy; the result of not conforming to requirements or, in other words, a potential task resulting from the existence of defects.

Procedural flow: synonymous with workflow.

Procedure: logical organization of people, materials, energy, equipment, and methods into repeatable work activities designed to produce a specified end result (work product); the repeatable and deterministic dimension of the holistic process model.

Process: a system in operation to produce an output of higher value than that of the sum of its inputs; the logical chain of customer-supplier relationships which drive work activities performed by people and other resource-consuming components to produce committed outcomes at specified points in time. The process comprises the customer-supplier communication protocol and the process flow (procedural flow). *See also* procedure.

Process capability: inherent ability of a process to perform consistently under given operating conditions; a measure of the inherent uniformity of the process.

Process control: the set of activities employed to detect and remove special causes of variation in order to maintain or restore stability (statistical control); process control leads to quality control.

Process flow: combined flow of materials and work products among activities, and the flow of related descriptive information. Synonymous with workflow.

Process improvement: set of activities employed to detect and remove common causes of variation in order to improve process capability; process improvement leads to quality improvement.

Process management: management approach comprising the three phases of process planning (design), process transfer (implementation planning and deployment), and process operations. The latter includes both commitment management and workflow management. *See also* process optimization.

Process map: symbolic representation of a business process in the form of a flow chart or diagram showing process inputs and outputs, the sequence of work activities, and flow of information within the process. *See also* mapping, process model.

Process model: a symbolic (mathematical, textual, or graphic) representation of the process as it currently exists ("as is") or as it is intended to be ("should be").

Process optimization: the major aspect of process management which concerns itself with maximizing the efficiency and productivity of the process — that is, with economic factors, while maximizing its effectiveness.

Process owner: designated executive or manager within the process who is accountable for the process output (outcome) and is empowered to manage the process and has the responsibility for its performance.

Process reengineering: fundamental analysis and radical redesign of a business process to achieve dramatic improvements in process performance (e.g., quality, cycle time, cost, and asset utilization).

Product: the output of any process, of both personal and general use. The term is related more to tangible goods, such as clothing, food, appliances, vehicles, office and data processing equipment, and so on. By extension, often used to include services and other similar offerings. *See also* offering and service.

Productivity: value added by the process divided by the value of the resources (including labor and capital) consumed. As a quantitative measure, it is the reciprocal of efficiency. *See also* efficiency.

Quality: conformance to requirements. The definition of quality includes (1) satisfaction features which result in customer satisfaction, and (2) freedom from deficiencies which result in the absence of customer dissatisfaction.

Quality control: totality of activities comprising the detection and correction of variation from established quality standards; the managerial process of maintaining or restoring the state of conformance to requirements.

Quality function deployment (QFD): comprehensive application of the orthogonal array technique for the systematic and successive deployment of customer needs in product and/or process development.

Quality management: totality of management actions for planning, controlling, and improving a state of conformance to requirements. *See also* total quality management.

Recipient: general and consistent definition of all those individuals or organizations for whom work products are directly provided. When a work product is acquired with some form of compensation in return, the recipient is known as a client. If the acquisition is for the direct use of the work product, the recipient is known as the user. The end user is one who receives the intended benefit of the work product. (End users of consumable products such as food, fuel, and clothing are known as consumers.)

Recomposition: reassembly or restoration of a process entity from its component subentities (tasks, activities, or subprocesses).

Reengineering: term frequently used to describe radical change of design, resulting in dramatic performance improvement. *See also* process reengineering.

Reliability: probability of a product or service performing its specified function under specified conditions, without failure, over a specified period of time.

Remedy: a change that can successfully remove or neutralize the cause(s) of a defect.

Requirement: a formal statement of a need and the expected manner in which it is to be met (the customer's view of the work product).

Scatter diagram: a graphic method of data representation for analytical purposes used to determine if a cause-and-effect relationship exists.

Sensor: a specialized device or means of detection or observation, designed to recognize the presence and intensity of certain phenomena and to convert them into information.

Service: activities performed for a customer, either related to goods and products (e.g., delivery, training, maintenance) or offered by themselves, such as electricity, transportation, education and training, and so on, to meet user requirements.

Simulation: the technique of observing and manipulating an artificial mechanism (model) that represents a real-world process which, for technical or economic reasons, is not suitable or available for direct experimentation.

Six Sigma: a statement of process capability; having a "Six Sigma process" means that its output will consistently achieve an average quality level of being 99.9996% defect free.

Software: collection of computer programs in symbolic representation.

Special cause: a source of variation in the process output that is unpredictable, unstable, or intermittent; also called assignable cause.

Specification: document containing a detailed description or enumeration of particulars; formal description of a work product and the intended manner of providing it; the supplier's view of the work product.

Stability (of process): a state of statistical control.

Stakeholder: a functional executive or manager who owns or controls tangible resources in a business process.

Strategy: a broad course of action, chosen from a number of alternatives, to accomplish a stated goal in the face of uncertainty.

Subprocess: a large collection of activities performed to transform inputs into outputs; a process that is part of, but subordinate to, a larger process; a single execution of all activities required to produce one process output; a single transaction or business cycle. *See also* activity and process.

Supplier: anyone who provides inputs to a process; individuals or organizations from whom work products originate, including agricultural producers, manufacturers, public utilities, educational institutions, etc. Suppliers who receive compensation for the inputs provided are known as vendors.

System: logical organization of people, methods, mechanisms, and automata deployed to accomplish a specified result (objective).

Task: smallest unit of repeatedly executable and measurable work to transform inputs into outputs; usually performed by one or more individuals.

Test: *see* appraisal.

Total quality management (TQM): term used to describe broad implementation of quality management company-wide; a customer-focused, unifying management approach encompassing every aspect of an organization's operations; also known as total quality control (TQC). *See also* quality management.

Ultimate user: the customer at the final destination of the product or service; the recipient of the intended benefit of the product or service. Synonymous with end user. *See also* user.

User: a customer who carries out positive and direct actions with respect to the product or service received (e.g., further processing or ultimate use). *See also* customer and ultimate user.

Variable: a data item which takes on values within some range with a certain frequency or pattern. Variables may be discrete — that is, limited in value to integer quantities (for example, the number of bolts produced in a manufacturing process); discrete variables relate to attribute data. Variables may also be continuous — that is, measured to any desired degree of accuracy (for example, the diameter of a shaft); continuous variables relate to variables data.

Variation: any nonconformance or deviation; differences among individual results or outputs of a process; generic term for defects, nonconformances, etc.

Work flow: sequence of procedural work steps, tasks, and activities and their interim work products within a process. *See also* procedural flow and process flow.

Work step: smallest definable and repeatedly executable unit of work to transform inputs into outputs; usually performed by one individual.

Endnotes

Endnotes

Introduction

1. IBM Advanced Business Institute, Palisades, NY, June 1998.
2. *The 1999 Conference on Knowledge Management and Organizational Learning,* The Conference Board, New York, April 15–16, 1999.
3. *The Economist,* September 19–25, 1998, p. 9.
4. Sweeney, D.M., Linking knowledge management and the world of *e*-business, in *The 1999 Conference on Knowledge Management and Organizational Learning,* The Conference Board, New York, April 15–16, 1999.
5. Computer age gains respect of economists, *The New York Times,* April 14, 1999.
6. *The Conceptual Foundations of Business Process Adaptability*: Designing and building a responsive and adaptive enterprise is like building a house. Its foundations must be solid and resting on the bedrock of sound theory and tested management concepts. The theoretical and conceptual foundations for this book have been laid by a few leading thinkers whose thought-provoking but always logical observations and propositions are particularly relevant to organization, governance, and business performance in the information age. The definition of business processes as the operational implementation of thermodynamically open systems is based on the post-World War II research and synthesis by von Bertalanffy (*General System Theory,* George Braziller, NY, 1968), whose principles of entropy and equifinality have been the cornerstones of all further study into adaptive systems ever since. The application of open system theory to social structures and, in particular, to corporate organization is primarily Ackoff's accomplishment (*The Democratic Corporation,* Oxford University Press, Cambridge, MA, 1994). His proposition that enterprises should be managed as systems is fundamental to the practical implementation of adaptive behavior both at the enterprise and business process level. A key characteristic of self-organizing systems is how they use their stability over time. Intermittent stability is the stepping stone to successively improved levels of organization which then lead to globally stable structures. Such a process remains stable at the macrolevel, while local disturbances exist and are tolerated; it allows for many levels of autonomy within itself. These multiple levels of autonomy represent, in the management

sense, multiple points of accountability. By accommodating these, the process is able to preserve its global stability with respect to the environment. Based on this behavior of self-organizing processes, it was Allan Scherr (A new approach to business processes, *IBM Systems Journal,* 32(1), 1993) who first introduced commitment and personal accountability as the management mechanisms for business processes in the face of change. In Scherr's model, commitments are the results of an ongoing dialogue between two parties: customer and supplier. Haeckel (see From "make and sell" to "sense and respond", *Management Review,* October 1992; *Adaptive Enterprise: Creating and Leading "Sense-and-Respond" Organizations,* Harvard Business Press, Cambridge, MA, 1999) extended Scherr's concept to enterprise governance. He argues that in transforming the traditional "make-and-sell" way of doing business into a highly responsive and adaptive "sense-and-respond" enterprise, the company-wide management of accountabilities is the only answer. Both Scherr and Haeckel recognize that key to the enterprise's adaptive behavior are its business processes through which strategies are executed and all operational business conducted. However, neither of them suggests a practical approach to systematic design and implementation. The purpose of this book, then, is to (1) briefly examine the characteristics and needs of the "sense-and-respond" enterprise and how it deals with unpredictable change; (2) show that "sense-and-respond" enterprises must be process centered; (3) examine what this means at the business process level; and (4) suggest a design approach for its manageable implementation. In this sense, the book is the "how-to" handbook of business executives, functional managers, and process designers for designing adaptable business processes as the basic building blocks of the process-centered, adaptive, "sense-and-respond" enterprise.

7. Hammer, M. and Champy, J., *Reengineering the Corporation,* Harper Business, New York, 1993.
8. Keen, P.G.W., *The Process Edge,* Harvard Business School Press, Cambridge, MA, 1997.

Chapter 1

1. The global, or at least international, movement of capital is not new. Since as early as the 13th century when Florentine merchants lent money to King Edward I of England to finance his wars, international capital has flowed across national boundaries and around the world in search for higher return. What is new is the enormous growth of capital flows. A good indicator of global capital flow is the amount of foreign currency bank deposits around the world, which surpassed $1 billion in 1961 but in little more than 30 years has exceeded $1 trillion (a thousandfold increase). This enormous pool of capital, literally sloshing from country to country, is looking for safe investments and higher profits. In the process it can and has brought great benefits to the global and certain individual economies; it can, however, also destabilize entire national or even regional economic systems.
2. Ackoff, R.L., *The Democratic Corporation,* Oxford University Press, New York, 1993, p. 23ff.
3. While it was Henry Ford who first introduced the moving assembly line into the automobile industry, he did not invent it. Records show that Venetian shipbuilders used the moving assembly line as early as the 16th century to construct their famous war galleons.

4. The term "organismic," introduced by Ackoff (see Ackoff, R.L., *The Democratic Corporation,* Oxford University Press, 1994) is a view of organizations that likens them to living organisms (e.g., the human body).

5. Hammer, M. and Champy, J., *Reengineering the Corporation,* Harper Business, New York, 1994, p. 29 (adapted with permission).

6. A partial explanation for the productivity improvements of the last decade, of course, may lie in low interest rates, but interest rates are staying low partially because the economic activities generated by the network economy are not necessitating an increase in the cost of capital.

Chapter 2

1. Peters, T.J. and Waterman, R.H., *In Search of Excellence,* Warner Books, New York, 1982.

2. Juran, J.M., *Juran on Leadership for Quality,* The Free Press, New York, 1989, p. 35.

3. Hammer, M. and Champy, J., *Reengineering the Corporation,* Harper Business, New York, 1993.

4. *State of Re-engineering Report 1994,* CSC Index, Cambridge, MA.

5. Hammer, M., Obliterate, don't automate, *Harvard Business Review,* July-August, 1990.

6. Keen, P.G.W., *The Process Edge,* Harvard Business School Press, Cambridge, MA, 1997, p. 1 ff.

7. Following Ackoff's approach, *transformation* is defined as discontinuous or fundamental change that addresses both function and structure — that is, *what* the organization does and *how* it does it. *Reformation* or *improvement,* on the other hand, only address one or the other and represent incremental or continuous change.

Chapter 3

1. For purposes of this discussion, we define value as something having the property of worth or desirability at a certain time. Value is highly subjective; the value of a product, service, or experience is never independent of those receiving or being affected by it, and therefore may be different for different recipients.

2. There are other categorizations: the two most frequently used recognize two (customer value or no value), or four (customer value, business value, operational necessity, and no value) categories. *Operational necessity* describes situations where work activities must be performed to maintain operational continuity in the face of known deficiencies, impediments, and the like.

3. *The Economist,* February 10, 1996, p. 61.

4. It is beyond the scope of this book to entertain a detailed discussion on economic value theory and related subjects. Suffice it to say, that economic value added (EVA) is defined

as the after-tax cash return on invested capital, less the cost of that capital. As business processes represent a substantial portion of an enterprise's total capital assets, EVA becomes an essential tool of business process management. Unlike reported paper profits and earnings per share, EVA correlates strongly with shareholder value (that is, the total market value of the company) and, by extension, with the enterprise's share of new value creation at the national or global level.

5. In this book, we will use the term *cost of nonconformance* as the expression for waste, because it is more consistent with the basic premise that the outcome of all business processes must meet agreed-to conditions of satisfaction — that is, *conform* to customer requirements.

6. The term *cost of quality* is well known and almost universally used, including its foreign language translations. In line with the previous note, however, the term *cost of satisfaction* would be more consistent and appropriate.

7. The role of the cost of quality and, by extension, of process cost is particularly important when an enterprise uses the balanced business scorecard (BBSC) as its overall performance measurement mechanism. Within the BBSC approach, the translatability and linkability of major measurement categories are essential to the method's success. The cost measurements related to process performance certainly meet this criterion.

8. This term was originally introduced by Steve Haeckel of IBM in his paper entitled "From 'Make and Sell' to 'Sense and Respond'" (*Management Review,* October 1992). It describes the ideal organization for effectively sensing, interpreting, and responding to unpredictable change. See also Bradley, S.P. and Nolan, R.L., *Sense and Respond,* Harvard Business School Press, Cambridge, MA, 1998.

9. See Ackoff, R.L., *The Democratic Corporation,* Oxford University Press, Cambridge, MA, 1994, p. 31ff.

Chapter 4

1. *The 1999 Conference on Knowledge Management and Organizational Learning,* The Conference Board, New York, April 15–16, 1999.

2. See Saint-Onge, H., Building the intellectual capital of the organization, in *The 1996 Strategic Management Conference of the Conference Board,* The Conference Board, New York, January 18, 1996.

3. Decisions, inasmuch as they result in commitments, add value through the EVA or IVA generated by the subsequent output of the business process that meets the conditions of satisfaction. When a decision does not result in a commitment or is explicit as not to make a commitment, its net added value (if any) may be realized over a longer period of time. This, then, becomes the arena for speculation and what-would-have-happened-if... hypothesizing.

4. Commitments and the process outcomes they cause to happen will most likely result in various forms of IVA, increasing structural capital (intellectual assets and the knowledge base), enhancing customer satisfaction and loyalty, and enriching human capital (new skills, lessons learned, experience, and knowledge of employees).

Chapter 5

1. The classic IPO model defines a deterministic process that is expected to be stable to perform in a predictable manner. Once underway, the process unfolds in a predictable sequence. In social process theory introduced later in this chapter and used through the balance of this book, the IPO definition becomes that of a *procedure.*

2. The following rules and attributes describe the workings and behavior of any process under consideration:*

   - When moving into one subject process, one will always find more, smaller processes. This is known as decomposition.
   - When moving out of one subject process, one will always enter more, larger processes. This is known as recomposition.
   - Inputs from the environment cross boundaries and enter subject processes. All inputs to any process are outputs of some other process.
   - Within the subject process, inputs interact with process components in a set of processing activities (also known as work activities).
   - The direction of process flow indicates the progression of events within the subject process. It is comprised of the following: (a) workflow, which is the sequence of procedural activities and the flow of materials and work products among activities, and (b) information flow, which is into, within, and out of the process. Information is made up of a descriptive component related to work and products, and a commissive component related to customer-supplier communications and, in particular, to commitments.
   - Processed inputs exit the subject process as outputs, crossing the boundary into the environment (e.g., into one or more customer processes). Outputs include tangible products or such intellectual results as decisions, agreements, and the like. All outputs of any process are inputs to some other process.
   - All processes function as either supplier (source of output) or customer (recipient or user of output or outcome) processes at any given time.
   - The output of the supplier process is called a product if it is a satisfactory input to the customer process.
   - Any output which is unacceptable to the customer process and the work and resources required to produce it is called waste. Because all outputs are inputs to some other process, waste may enter another process by default or remain in the supplier process (e.g., waste collects in the supplier process or in the customer process, or both).
   - If the outputs of the supplier process do not satisfy the environment (that is, the customer), the inputs into that process will diminish or cease.
   - Outcomes are the effect that outputs have on subsequent processes, sometimes also called downstream processes. A single execution of the subject process may produce one output, but several outcomes.
   - One can see one's output from within the subject process. To see outcomes, however, one has to move out of the subject process and into the downstream processes affected by one's output. One way of doing this is through the introduction of

---

* For an interesting, systems-oriented discussion on this subject, see Cusins, P., Understanding quality through systems thinking, *The TQM Magazine,* Vol. 6, No. 5 (October), MCB University Press, Bradford, U.K., 1994. (Adapted here with permission.)

extended feedback loops or learning loops (described in more detail in the Process Feedback section).

- Beneficial or adverse outcomes may sometimes be seen only by looking at affected processes which may be a number of steps removed from the subject process. Such outcomes may take considerable time to become visible.

3. The laws of process quality include:*

- Quality is a boundary judgement made at the interface of two processes.
- In total quality management (TQM), effective boundary judgments are made at every process level within an existing organization; however, TQM does not propose to change the organization itself. It is meant to maintain stability or eliminate deficiencies, not address adaptability to change and its future consequences.
- An output is considered product or waste based on judgment about its quality; that is, whether or not it meets previously agreed-to conditions of satisfaction.
- Neither good quality nor poor quality will be noticed, provided there is a complete match in perception between what the customers are getting and what they are expecting.
- Static quality factors are those ensuring freedom from deficiencies. They are not noticed when present; however, their absence will create an image of poor quality and result in customer dissatisfaction. They are general and common to all customers, and they are situation independent.
- Dynamic quality factors are known as satisfaction features. They are not noticed when absent, but their addition will create an image of high quality and result in customer satisfaction. They are individual and unique to each customer, and they are situation dependent.

4. An open system is defined as a system in exchange of matter or energy with its environment. All social processes which include humans are in fact open systems. As classic system theory was expanded to open systems, two basic principles of open systems have emerged:

- The principle of equifinality states that the same final state of an open system may be reached from different initial conditions and through various alterations of the process. This principle is the basis for the phenomena of self-organization and adaptability.
- The principle of entropy states that in open systems a transition toward higher order and organization is possible along with an import (negative change) of entropy. The basic notion in communication theory is that of information. By extension, information representing change can be viewed as negative entropy — the input required to trigger the desired process response. This means that open systems containing humans can maintain themselves in a steady state and then move to a higher level of organization. This principle is the basis for the phenomenon of responding to change with an improved level of capability.

Some of the key results of post-World War II research in this field are presented by Ludwig von Bertalanffy in his excellent review *General System Theory* (George Braziller, New York, 1968).

---

* Adapted from Cusins, P., *The TQM Magazine,* Vol. 6, No. 5 (October), MCB University Press, Bradford, U.K., 1994. With permission.

5. The holistic definition of business processes is rooted in the original work of Fernando Flores (Management and Communication in the Office of the Future, Ph.D. Dissertation, University of California, Berkeley, 1962) and others, who established the view of business processes as a series of formalized conversations or dialogues. Allan Scherr introduced the customer-supplier relationship as the carrier of these conversations and the commitment as its key result (Scherr, A.L., A new approach to business processes, *IBM Systems Journal,* 32(1), 1993).

Chapter 6

1. The source of this definition of transformation is Russell Ackoff, who likes to remind his audiences that improvement is only reformation, a change in either the function or the structure of an existing system entity, but not both.
2. *The Process-Centered Organization: Principles and Design,* Hammer and Company, Cambridge, MA, 1996.
3. Commitments are always understood to be made by suppliers to customers.
4. This assumption has its roots in the basic system-oriented approach to corporate governance, best articulated by Ackoff and Haeckel in the mid-1990s (Ackoff, R.L., *The Democratic Corporation,* Oxford University Press, New York, 1994; Haeckel, S.H., Adaptive enterprise design: the sense-and-respond model, *Planning Review,* 23(3), 1995.)
5. In this text, the term "interface" means a boundary between two processes (or two subprocesses, etc.) across which work products or information flow. A relationship is an interface defined by the interactions of two parties (usually empowered people) playing the respective roles of customer and supplier — hence, customer-supplier relationship.
6. Grant Thornton Survey of American Manufacturers, Chicago, IL, 1993.
7. Seventh Annual Survey, CSC Index, Cambridge, MA, 1994.
8. Haeckel distinguishes two levels of adaptability (*Adaptive Enterprise: Creating and Leading "Sense-and-Respond" Organizations,* Harvard Business Press, Cambridge, MA, 1999). One is the enterprise itself; its adaptability is treated as a leadership and policy issue for which senior management is accountable. Adaptability *within* the enterprise is at the level of individual business processes, which is the primary focus of this book.
9. This adaptive loop for processes is an extension of the well-known "observation-orientation-decision-action" (OODA) loop used by the U.S. Air Force, or the "sense-interpret-decide-act" (SIDA) loop, to specifically include the step of commitment. For a concise description, see Haeckel, S.H. and Nolan, R.L., Managing by wire, *Harvard Business Review,* September-October, 1993.
10. See Haeckel, S.H., Adaptive enterprise design: the sense-and-respond model, *Planning Review,* 23(3), 1995.
11. Haeckel, S.H., *Commitment Management,* IBM Advanced Business Institute, Palisades, NY, February, 1996.
12. For the definition of accountability as related to business process management, see Chapters 8 and 16.

Chapter 7

1. For purposes of this discussion, we will follow Ackoff's definition of organizational development: the increasing ability to satisfy the needs and expectations of the enterprise within the framework of a stated purpose. (For further reading on this subject, see Ackoff, R.L., *The Democratic Corporation,* Oxford University Press, New York, 1994.)
2. Accelerating and unpredictable change is also known as discontinuity. By definition, a discontinuity occurs when any or all of the following conditions exist: (a) known variables exceed their range of predictability, (b) new variables are introduced, and (c) internal (process) reaction time exceeds the time interval at which changes in the environment occur. (For further reading, see Haeckel, S.H., *Adaptive Enterprise: Creating and Leading "Sense-and-Respond" Organizations,* Harvard Business Press, Cambridge, MA, 1999.)
3. Haeckel, S.H. and Nolan, R.L., Managing by wire, *Harvard Business Review,* September-October 1993, page 122. [Adapted with permission.]
4. This conclusion and the following paragraphs are based on Haeckel's interpretation of Berkeley Professoir Rashi Glazer's study of businesses operating in environments of increasing information intensity. For further reading on this subject, see Haeckel, S.H., *Adaptive Enterprise: Creating and Leading "Sense-and-Respond" Organizations,* Harvard Business Press, Cambridge, MA, 1999; Glazer, R., Marketing in an information-intensive environment, *Journal of Marketing,* October, 1991.
5. For a detailed description of the formal mechanism for these interactions, known as the customer-supplier dialogue, see Chapter 3.
6. See Ackoff, R.L., *The Democratic Corporation,* Oxford University Press, New York, 1994.
7. See Gertz, D. and Baptista, I., *Grow To Be Great,* Free Press, New York, 1996.
8. Hamel, G., How killers count, *Fortune,* June 23, 1997, p. 74.

Chapter 8

1. Juran calls this to be in a state of self-control. (See Juran, J.M., *Juran on Leadership for Quality,* The Free Press, New York, 1989, p. 147.)
2. The term "empowerment" is not universally liked. In a recent interview, Peter Drucker declared, "I have never used the word empowerment and never will." He then reminded his readers that he has always talked of responsibility and "only of responsibility." As an explanation, he suggested that empowerment implies an organization is based on power — in which case it does not matter where the power resides: at the top or at the bottom. (See Galagan, P.A., Peter Drucker, *Training and Development,* September, 1998.) This book has adopted the term because it is widely used and probably not entirely with the same intent as what Drucker suggests. In our treatment of the subject, commitment and accountability together correspond to Drucker's notion of responsibility, because they also presume authority.
3. Haeckel, S.H., *Adaptive Enterprise: Creating and Leading "Sense-and-Respond" Organizations,* Harvard Business Press, Cambridge, MA, 1999.

Chapter 9

1. Several authors, among them Michael Hammer, consider such a council optional. In our experience, the formal and continuing focus and involvement of senior management are critical. While this focus may be achieved in different ways, regular attention given by a body of key policymakers to matters of process centering and reengineering is the most effective as well as practical solution.
2. Hammer, M. and Champy, J., *Reengineering the Corporation,* Harper Business, New York, 1994, p. 109.

Chapter 10

1. These conclusions are based on Glazer's definition and Haeckel's related interpretation of information intensity. (See Glazer, R., Marketing in an information-intensive environment, *Journal of Marketing,* October, 1991; Haeckel, S.H., *Adaptive Enterprise: Creating and Leading "Sense-and-Respond" Organizations,* Harvard Business Press, New York, 1999.)
2. In this context, it is important to consider the very unique economic properties of information as an asset: (a) it is infinitely divisible (that is, one can distribute it endlessly without losing it), (b) it does not depreciate, and (c) its value increases with sharing and usage. These attributes are entirely different from those of traditional economic assets: material, energy, capital, land, and labor, all of which may be appropriated (one cannot give them away without giving them up). They also depreciate with time and tend to lose their value with progressive usage.
3. Notwithstanding the slow but steady advance of computer-based artificial intelligence which is capable of limited interpretations within a specified set of parameters.
4. See Blickstein, S., Does training pay off?, *Across the Board,* XXXIII(6), 1996.
5. This is also known as Say's law; the observation that supply creates its own demand dates back to the 19th-century French economist, Jean Baptiste Say.

Chapter 11

1. See Garvin, D.A., The processes of organization and management, *Sloan Management Review,* Summer 1998, p. 33ff.
2. See Davenport, T.H., *Process Innovation: Reengineering Work through Information Technology,* Harvard Business School Press, New York, 1993, p. 28.
3. Peter Keen's four categories of identity, priority, background, and mandated processes are based on their salience (synonymous with the term "relevance" as used in this book). See Keen, P.G.W., *The Process Edge,* Harvard Business School Press, New York, 1997, p. 25ff.
4. Based on where and by whom they are carried out, Garvin divides the multitude of processes into two major categories: (a) organizational processes (comprising work,

behavioral, and change processes) implemented across the organization, and (b) managerial processes (including direction-setting, negotiating and selling, and monitoring and control processes), essentially reserved for managerial action. See Garvin, D.A., The processes of organization and management, *Sloan Management Review,* Summer, 1998.

5. The International Benchmarking Clearinghouse, in partnership with Arthur Andersen & Co., has described 71 major processes and an equal number of identifiable subprocesses in two overarching categories: operating processes and management and support processes. See *Process Classification Framework,* APQC International Benchmarking Clearinghouse, Houston, TX, 1993.

6. Evolutionary processes include: (a) natural evolution, such as cosmic or geological change, radioactive decay, the life cycle of living organisms, economic development, the life cycle of human organizations and societies, and the like; (b) planned change, induced by humans, primarily to modify the course of natural evolution, such as economic or social development, and the environment.

7. Coordination, in this context, means linking commitments to achieve a defined final outcome.

8. At the time that the performance goals for on-time delivery and response to customer inquiries were established, they represented unprecedented levels in the industry. For this reason, instead of being demotivators, they proved to be strong drivers for the process team and all others involved to design an order-fulfillment process whose performance would exceed all known competing processes at the time and into the foreseeable future.

9. *Webster's New Collegiate Dictionary,* 2nd ed., Merriam, New York, 1958.

Chapter 12

1. See Ackoff, R.L., *The Democratic Corporation,* Oxford University Press, New York, 1994, p. 22ff.

2. Work activities can be pre-planned, in which case they are considered procedural, or they can be *ad hoc,* created as a new capability in response to a changed or new requirement.

3. A capability in this context is defined as the procedures and resources that produce outcomes contributing to the overall output of the business process. In practical terms, such capabilities are usually viewed as equivalent to subprocesses within the business process.

4. See Winograd, T. and Flores, F., *Understanding Computers and Cognition: A New Foundation for Design,* Ablex Publishing, Stamford, CT, 1986.

5. See Scherr, A.L., A new approach to business processes, *IBM Systems Journal,* 32(1), 1990.

6. See Haeckel, S.H., *Commitment Management,* IBM Advanced Business Institute, Palisades, NY, 1996.

7. A good overview of transaction-cost economics is given by Haeckel, based on the original work by British economist Ronald Coase in the late 1930s. In his paper published in 1937, Coase explored how firms decide which aspects of their operations to handle internally

(processing) and which to buy (transactions) as being the very foundations of operational management. (See Coase, R., The nature of the firm, *Economica,* 4, 386–405, 1937; Haeckel, S.H., *Adaptive Enterprise: Creating and Leading "Sense-and-Respond" Organizations,* Harvard Business Press, Cambridge, MA, 1999.)

8. Information may not always be renewable at the source, or in terms of its currency; however, as discussed in Chapter 8, it is infinitely replicable.
9. The process measurement plan documents the entire set of measurement points in and around the business process.

Chapter 13

1. The various enterprise-wide teams and councils, such as the transformation council or steering committee, are organizational means, not a structural part of the process design.
2. This concept is based on the research on dissipative structures by Jantsch and others. (See Jantsch, E., *The Self-Organizing Universe,* Pergamon Press, Elmsford, NY, 1980, p. 30ff.)
3. Jantsch, E., *The Self-Organizing Universe,* Pergamon Press, 1980, Elmsford, NY, p. 141.
4. Scherr (1993) refers to it as complete process definition, while Center and Henry (1993) and several other authors use the term "workflow paradigm." Given that the term "workflow" has a procedural connotation (just the opposite of what we are trying to suggest here), in this book we have decided in favor of using the adjective "holistic" when referring to the two-dimensional, integrated process model or its management.
5. This approach is presented as a model, because it is possible to present a formal description of how it works.
6. The names of the four phases and of the related communication acts are based on and adapted from the theoretical work of Flores and Winograd (1986), practical application by Scherr (1993), and several, primarily IBM, authors following his approach.
7. Flores and Winograd (1986) and several other authors following in their footsteps call these "speech acts." Scherr (1993) and Haeckel (1999) remind their audiences that, as they contain at least implied commitments, the proper term is "commissive acts." In this book, we have chosen to use the term "communication act," being well aware that the term has a broader meaning, in that communications may be descriptive, informative, as well as commissive.
8. This rule is one of the cornerstones of the dialogue concept. Its corollary is that commitments that cannot be fulfilled must be renegotiated or else the committing party must withdraw. Sadly, it is the rule most frequently violated or ignored in business life; we all know of pending applications, unanswered letters or telephone calls, missed appointments, broken commitments, unreported changes or obstacles, and so on.
9. The adaptive loop is also known as the feedback loop, the learning loop, the OODA (observe-orient-decide-act) loop, or the SIDA (sense-interpret-decide-act) loop. An adaptive loop has four functional steps (or phases): (a) observation or sensing — recognize and capture information from both within the process and the external environment; (b) orientation or identification — interpret that information within a

specific context; (c) decision — select a response based on the interpretation; and (d) action — execute the response.

10. See Center, K. and Henry, S., A new paradigm for business processes, in *Proc. of the Workflow Conference on Business Process Technology,* May 3–5, 1993, Boston, MA, sponsored by the Workflow Institute, Alameda, CA.

Chapter 14

1. Sun Tzu, *The Art of War,* Shambala Pocket Classics, Boston, MA, 1991.
2. For this reason, continuous workflow is also defined as one in which there are only value-adding work steps. This, in turn, leads to the time-based measurement of theoretical cycle time: the length of time consumed by only value-adding work steps in the process.
3. The definition of a choke point is a hand-off point where the flow of process resources is impeded such that it causes unacceptable delays against stated time-based performance (schedule or cycle time) objectives. Choke points are sometimes referred to as bottlenecks or logjams.
4. *Best Practices in Business Process Reengineering,* Oxford Associates/BPRC, Bethesda, MD, 1993.
5. Confidential survey of key client organizations by Juran Institute, Wilton, CT, 1998.
6. See Dyson, E., *Release 1.0,* Edventure Holdings, June 1988 and November 1990; *Release 2.0,* Broadway Books, NY, 1997.
7. A discussion of the selection criteria for ERP systems is beyond the scope of this book. Suffice it to list here the top four worldwide providers of ERP solution software in descending order of 1997 worldwide software revenues: SAP, Oracle, PeopleSoft, and Baan.
8. Juran Institute's *Business Process Quality Management® (BPQM)* method was used under license in the high-level process redesign.
9. Business Transformation Design (BTD) of Stamford, CT; Action Technologies of Alameda, CA; and a handful of other consultants offer descriptive software that supports a form of the holistic model (workflow paradigm in their terminology).
10. Examples are Intellicorp's *Livemodel,* Visio Corporation's *VISIO,* and IDS Scheer's *ARIS* modelers.

Chapter 15

1. Haeckel, S.H., *Adaptive Enterprise: Creating and Leading "Sense-and-Respond" Organizations,* Harvard Business Press, Cambridge, MA, 1999.
2. Pall, G.A., Holistic business process model for managing workflow, commitments and accountabilities, *Juran Institute Technical Report,* Juran Institute, Wilton, CT, 1996.

Chapter 16

1. The principle of equifinality states that the same final state of an open system may be reached from different initial conditions and through various alterations of the process. This principle is the basis for the phenomena of self-organization and adaptability.
2. Tasks offer the practical base for activity-based costing (ABC), the method for capturing the resource consumption of the process in financial terms. In this sense, the term "activity-based" is a misnomer, as it is really tasks that serve as the basis for the costing work. For this reason, many organizations now use the designations task-based costing, task-based billing, task-based design, and the like.
3. Humphrey, W.S., *Introduction to the Personal Software Process,* Addison-Wesley, Reading, MA, 1997.
4. Scherr, A.L., A new approach to business processes, *IBM Systems Journal,* 32(1), 91, 1993.
5. Haeckel, S.H., Adaptive enterprise design: the sense-and-respond model, *The Planning Review,* 23(3), 1995.
6. For the definition of discontinuity, see Endnote 2 in Chapter 7.
7. Russell Ackoff likes to point out that process improvement is a form of reformation as opposed to transformation.
8. See Ackoff, R.L., *Creating the Corporate Future,* John Wiley & Sons, New York, 1981.
9. In the non-English speaking countries of the world, and particularly in Europe, the year 2000 is referred to as MM, which is 2000 in Roman numerals. The "Y2K problem" then becomes the "MM problem."
10. Too late to defuse millennium bomb, software group warns, *Financial Times,* November 10, 1998, p. 1.

Chapter 17

1. *Business Process Quality Management,* 2nd ed., Juran Institute, Wilton, CT, 1993.
2. Juran, J.M., *Juran on Leadership for Quality,* The Free Press, New York, 1989, p. 316.
3. For the most frequently used project management tools, see the Appendix.

Chapter 18

1. These subsystems usually include (a) performance management (planning and evaluation), (b) professional development (training and career management), (c) rewards and recognition (compensation and benefits), and (d) employment (hiring, termination, and retirement). In some organizations, a fifth subsystem may be dedicated to the management of benefits during and after employment.

2. Amazing as it may sound, the third approach has been followed by a number of companies where, upon completion of the first two phases of process reengineering (planning and transfer), functional management prevailed and succeeded in reasserting control over the process management infrastructure. This outcome negates much of the value of a cross-functional design effort and, because of the absence of accountability, will ensure a complete lack of adaptability in the face of future change.

3. The role of the process owner, *Phoenix Executive Summary,* Hammer and Company, Cambridge, MA, April 1995.

4. By now, CMM has a vast literature. For basic reading, three texts can be recommended: *The Capability Maturity Model: Guidelines for Improving the Software Process,* Carnegie Mellon University, Software Engineering Institute, Addison-Wesley, Reading, MA, 1994; Humphrey, W.S., *A Discipline for Software Engineering,* Addison-Wesley, Reading, MA, 1995; Raynus, J., *Software Process Improvement with CMM,* Artech House, Boston, MA, 1999.

Bibliography

Bibliography

Ackoff, R.L., *The Democratic Corporation,* Oxford University Press, New York, 1994.

Andrews, D.C. and Stalick, S.K., *Business Reengineering,* Yourdon Press, Upper Saddle River, NJ, 1994.

Argyris, C. and Schön, D.A., *Organizational Learning II,* Addison-Wesley, Reading, MA, 1996.

Bertalanffy, L. von, *General System Theory,* George Braziller, New York, 1968.

Bolman, L.G. and Deal, T.E., *Reframing Organizations,* Jossey-Bass, San Francisco, CA, 1997.

Bradley, S.P. and Nolan, R.L., *Sense and Respond,* Harvard Business School Press, Cambridge, MA, 1998.

Browning, J., The power of process redesign, *McKinsey Quarterly,* 1(1), 1993.

The Capability Maturity Model: Guidelines for Improving the Software Process, Software Engineering Institute, Addison-Wesley, Reading, MA, 1994.

Center, K. and Henry, S., A new paradigm for business processes, in *Proc. of the Workflow Conference on Business Process Technology,* May 3–5, 1993, Boston, MA, sponsored by the Workflow Institute, Alameda, CA.

Champy, J., *Reengineering Management: The Mandate for New Leadership,* Harper Business, New York, 1995.

Collins, J. and Porras, J., *Built To Last,* Harper Business, New York, 1994.

Davenport, T.H., *Process Innovation,* Harvard Business School Press, Cambridge, MA, 1993.

Flores, F. and Winograd, T., *Understanding Computers and Cognition: A New Foundation for Design,* Ablex Publishing, Stamford, CT, 1986.

Garvin, D.A., Leveraging processes for strategic advantage, *Harvard Business Review,* September-October, 1995.

Garvin, D.A., The processes of organization and management, *Sloan Management Review,* Summer, 1998.

Godfrey, A.B., Future trends: expansion of quality management concepts, methods and tools to all industries, *The Quality Observer,* 6(9), 1997.

Goldratt, E.M., *Theory of Constraints,* North River Press, Croton-on-Hudson, NY, 1990.

Gouillart, F.J. and Kelly, J.N., *Transforming the Organization,* McGraw-Hill, New York, 1995.

Haeckel, S.H., From "make and sell" to "sense and respond", *Management Review,* October, 1992.

Haeckel, S.H., *Adaptive Enterprise: Creating and Leading Sense-and-Respond Organizations,* Harvard Business Press, Cambridge, MA, 1999.

Haeckel, S.H. and Nolan, R.L., Managing by wire, *Harvard Business Review,* September-October, 1993.

Hall, G. et al., How to make reengineering really work, *Harvard Business Review,* November-December, 1993.

Hammer, M., Reengineering work: don't automate, obliterate, *Harvard Business Review,* July-August, 1990.

Hammer, M., *Beyond Reengineering,* Harper Business, New York, 1996.

Hammer, M. and Champy, J., *Reengineering the Corporation,* Harper Business, New York, 1993.

Hammer, M. and Stanton, S., *The Reengineering Revolution,* Harper Business, New York, 1995.

Harbour, J.L., *Cycle Time Reduction,* Quality Resources, NY, 1996.

Harrington, H.J., *Business Process Improvement,* McGraw-Hill, New York, 1991.

Humphrey, W.S., *A Discipline for Software Engineering,* Addison-Wesley, Reading, MA, 1995.

Humphrey, W.S., *Introduction to the Personal Software Process,* Addison-Wesley, Reading, MA, 1997.

Hunt, V.D., *Process Mapping: How To Reengineer Your Business Processes,* John Wiley & Sons, New York, 1996.

Johann, B., *Designing Cross-Functional Business Processes,* Jossey-Bass, San Francisco, CA, 1995.

Juran, J.M., *Juran on Leadership for Quality,* The Free Press, New York, 1989.

Juran, J.M., *Juran on Quality by Design,* The Free Press, New York, 1994.

Kane, E. J., IBM's quality focus on the business process, *Quality Progress,* Vol. 19, 1986.

Keen, P.G.W., *The Process Edge,* Harvard Business School Press, Cambridge, MA, 1997.

Kotter, J.P., *A Force for Change,* The Free Press, New York, 1990.

Kotter, J.P., *Leading Change,* Harvard Business School, Cambridge, MA, 1995.

Leonard, D., *Wellsprings of Knowledge,* Harvard Business School Press, Cambridge, MA, 1995.

Lockamy, A. and Cox, J.F., *Reengineering Performance Measurement,* Irwin, Chicago, IL, 1995.

Lowenthal, J.N., *Reengineering the Organization,* ASQC Quality Press, Milwaukee, WI, 1994.

Manganelli, R.L. and Klein, M.M., *The Reengineering Handbook,* American Management Association, 1994.

Melan, E., *Process Management,* McGraw-Hill, New York, 1993.

Norris, G. et al., *SAP: An Executive's Comprehensive Guide,* John Wiley & Sons, New York, 1998.

Pall, G.A., *Quality Process Management,* Prentice-Hall, Englewood Cliffs, NJ, 1987.

Poirier, C.C. and Reiter, S.E., *Supply Chain Optimization,* Berrett-Kohler, San Francisco, CA, 1996.

Prigogine, I. and Stengers, I., *Order Out of Chaos,* Bantam Books, New York, 1984.

Raynus, J., *Software Process Improvement with CMM,* Artech House, Boston, MA, 1999.

Re-engineering reviewed, *The Economist,* July 15, 1994.

Roberts, L., *Process Reengineering,* ASQC Quality Press, Milwaukee, WI, 1995.

Robson, G.D., *Continuous Process Improvement,* The Free Press, New York, 1991.

Royce, W., *Software Project Management,* Addison-Wesley, Reading, MA, 1998.

Rummler, G.A. and Brache, A.P., *Improving Performance: How To Manage the White Space on the Organization Chart,* Jossey-Bass, San Francisco, CA, 1995.

Scherr, A.L., A new approach to business processes, *IBM Systems Journal,* 32(1), 1993.

Senge, P., *The Fifth Discipline,* Doubleday, New York, 1990.

Shores, R., *Reengineering the Factory,* ASQC Quality Press, Milwaukee, WI, 1995.

Stewart, T.A., Reengineering: the hot new managing tool, *Fortune,* August 23, 1993.

Stewart, T.A., *Intellectual Capital,* Doubleday, New York, 1997.

Wheatley, M., *Leadership and the New Science,* Berrett-Koehler, San Francisco, CA, 1992.

Whiteley, R. and Hessan, D., *Customer-Centered Growth,* Addison-Wesley, Reading, MA, 1996.

Wind, J.Y. and Main, J., *Driving Change,* The Free Press, New York, 1997.

The wonders of workflow, *The Economist,* December 11, 1993.

Index

Index

A

B

C

D

N

O

Q

R

S

T